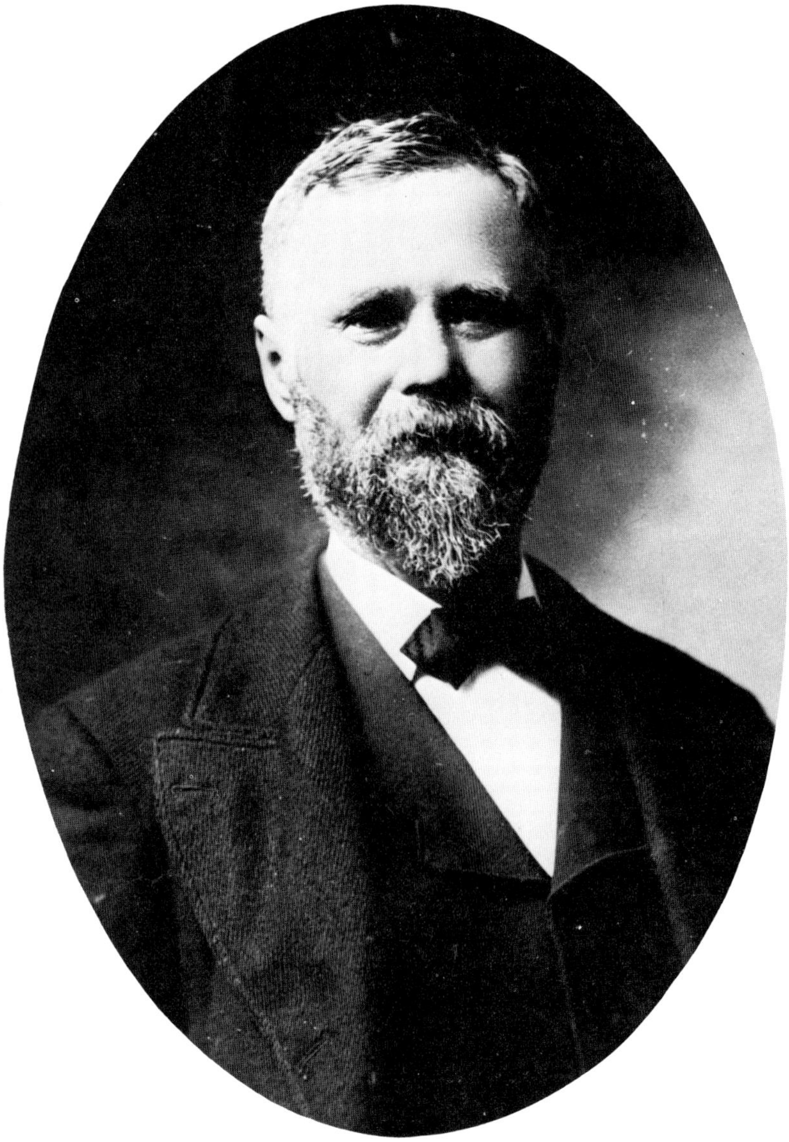

Secure the Shadow

Lachlan McLean, Colorado Mining Photographer

Secure the Shadow

Lachlan McLean
Colorado Mining Photographer

Duane A. Smith
Hank Wieler

Colorado School of Mines Press, Golden, Colorado

Guy T. McBride, Jr., President
Colorado School of Mines, Golden, Colorado 80401
Printed in the United States of America
Published by Colorado School of Mines Press

Library of Congress Cataloging in Publication Data

Smith, Duane A.
Secure the Shadow.
Bibliography: p.
Includes index.
1. Photography—Colorado—History. 2. Gold mines and mining—Colorado—History. 3. Silver mines and mining—Colorado—History. 4. McLean, Lachlan, 1843-1929. I. Wieler, Hank, joint author. II. Title.
TR24.C6S64 770'.92'4 80-10693
ISBN 0-918062-09-8

To
Laralee

Contents

Illustrations

Preface

The 1,556 glass negatives stored in the Western Historical Collections of the University of Colorado had not created much interest over the years, except when an occasional move revealed their startling weight, or when they required dusting now and then. The heritage of Lachlan McLean was nearly forgotten until Hank Wieler came along. At that point a story began to unfold, eventually becoming this volume.

Many individuals helped along the way, and deepest thanks are extended to them all. Their efforts are sincerely appreciated. George Kupfner, Director of the University of Colorado Photographic Services, hand printed all the photographs (some of them rather the worse for wear) from the original negatives. Jack Brennan, curator of the collections, encouraged and aided in many ways, beginning back in the midseventies.

Cassandra Volpe was always helpful, and Cathy Conrad again typed the manuscript from the rough pages given to her. Allan Bird provided a professional commentary on mining equipment, Liston Leyendecker checked for historical errors, and Gay Smith lent her skilled editorial touch and masterful typing. Jon Raese, director of publications for the Colorado School of Mines Press, was the glue which held this all together, and it was his support that made this book possible.

Prologue

The Quest

Hank Wieler

In the 1930s, the Norlin Library Western Historical Collections of the University of Colorado at Boulder acquired a collection of glass negatives, lantern slides, and stereoptican views. Nothing was recorded about the gift except that it had been produced by L. McLean, a prominent photographer in Idaho Springs, Colorado, for over 40 years. The negatives, for the most part, were in good condition and of excellent quality. The subjects ranged from mountain scenes, portraits, interesting events, down to a silhouette of a frog taken by the rays emitting from radioactive ore.

McLean, a photographer who started his career back in the days of the first wet plates, continued to work through the era of dry plates and into the modern age of roll film. There were both wet and dry plates in the collection, indicating he had begun to photograph the Rocky Mountains before dry plates came into general use.

The photographer of today cannot comprehend the hard work and the many problems that confronted the photographer of a hundred years ago. In the 1870s, taking a picture in the field was not a simple project. Photographs were made by using the "wet plate process"; each glass plate had to be prepared on the scene just before the picture was taken. This was difficult enough in the cities, but in the mountains all problems were magnified. Then, in 1880, George Eastman improved the dry plates, coming up with a glass plate that had the photographic emulsion already dried on the glass and packaged in a light-proof container.

Photography in those days was much more than simply pointing a camera at the subject, focusing, and pushing the trigger. The pioneer photographer first had to set up his heavy camera and tripod, carefully focus on the subject, and then set up his dark tent. Inside he placed the bottles of collodion, developer, fixing solutions, trays, and water. Colorado's early photographers made many prints that are fine shots even by today's standards. When one remembers the difficulties they had in doing their work, the finished product is even more remarkable.

With the introduction of the dry plate, the need for a portable darkroom was eliminated. However, if the photographer wished to photograph underground mining activity, he had to resort to the use of flash powder. Magnesium powder was placed in a hand-held trough and ignited to furnish the necessary light, a process that held an element of danger. If too much powder were used, it would explode rather than burn. There was always a risk of fire; underground, the added danger of flammable gas in some mines had to be recognized. After the flash exposure, the huge amount of dense white smoke it produced meant that everyone had to leave until the air cleared.

That was photography a hundred years ago. Considering the amount of work and skill involved, this collection is even more intriguing. Its creator, Lachlan

McLean, was obviously a photographer interested in more than just wedding pictures. But at the start of this quest, nothing of this man was known except his last name and first initial. Surely a businessman who lived so many years in one area should have left more of a record than just his work. It was only after 3 months of searching that evidence of his existence finally began to emerge.

The initial search was frustrating. Old-timers in Idaho Springs vaguely recalled L. McLean, and the *Colorado State Business Directory* listed him as a photographer in Georgetown, Colorado, in 1877. A later directory located his shop on Alpine Street, Georgetown, and a McLean negative of Alpine Street showed a sign advertising a photographer. Finding Alpine Street, however, was not an easy job. No one in town knew where it was, had been, or what the name might have been changed to. Finally, at the Clear Creek County Courthouse, a helpful clerk resurrected an 1876 map and there was Alpine Street, with a penciled notation that its name had been changed to Sixth Street.

An inspection of the site found the little two-story building next to a vacant lot. Though much the worse for wear and ready to fall down, it was still there—the same building in the photograph. But here, when the trail was just beginning to unfold, it was lost. Georgetown did not remember him. According to the *Colorado State Business Directory,* he had a shop in Georgetown only until 1881, a time too short for lasting impressions. He was not listed again until 1884, this time in Golden, Colorado. Apparently, he moved once more, and in 1885 the directory listed a shop in Idaho Springs. His last entry was in 1927.

More research in Georgetown produced nothing of value. The county and city records showed that in 1880 he had paid some taxes under the name of L. McLean. His first name remained elusive.

The voluminous western collection of the Denver Public Library was dredged, with no results. Genealogy and morgue files yielded nothing, but the library did have several pictures that were prints of McLean negatives. At last, it seemed, progress was being made. Unfortunately, this lead turned into a dead end, too. The pictures had been gifts, parts of old collections. They were tantalizing but undecipherable as to background.

The Colorado Historical Society would surely produce something about a photographer whose career spanned half a century. Again, a blank was drawn. To make matters worse, Norlin Library at the University of Colorado was uncertain as to how it had acquired the collection. One could be expected to wonder whether there ever was such a man as L. McLean, but Norlin Library offered all those beautiful negatives, many stamped L. McLean and all packed in sturdy wooden boxes, as proof that he actually existed.

Friends in Central City suggested names of several people residing in Idaho Springs who were interested in the history of the region, and they in turn led to Marjorie Bell. She was too young to remember him personally, but arranged a meeting with several friends who did remember him. After seeing his prints, they recalled the location shown in the photographs but could not bring to mind his first name. They did think he died in the late 1920s. Since he had been listed in the directory of 1927, that narrowed down the time of death, if they were right.

This same meeting elicited the fact that McLean had several children, one of whom had attended the University of Colorado and studied history under a long-gone history professor. This professor had made gifts of materials to the then infant archives. This possibly was the source of the collection, but a search of the library records drew another blank. In the 1930s, gifts had been only briefly recorded, not catalogued. Besides, 40 years ago a collection of glass negatives did not arouse exceptional interest, but the tide was turning.

At the Denver Public Library, a researcher overheard the conversation about early Idaho Springs and introduced himself as Paul Stewart, a Black historian of Denver. Stewart's interest in early Black inhabitants of Clear Creek and Gilpin counties led to a promise that he would receive any information about early pictures of Blacks that might be found in this quest. He was to play a role in the uncovering of McLean's past.

Meanwhile, back in Boulder, news was waiting. An assayer from Idaho Springs had visited the Western collections checking on historical photographs, and the archives staff, Sandra Volpe and Doris Mitterling, pounced on him about McLean. He did not remember the photographer very clearly but said his mother, now 95 years old and "sharp as a tack," had had her wedding photo taken by him. Thus George Treder entered the picture. More and more people were becoming involved in the McLean quest. The fact that there was no easy approach to information was challenging.

A phone call to Idaho Springs produced a meeting with Mrs. Treder. She remembered McLean as a man with a beard, who was not young when she knew him. The Treders did affirm that McLean had several children. Another trip to City Hall in Idaho Springs revealed no cemetery records, except for his children. Mrs. Treder remembered going to a McLean funeral service but thought it might have been for one of his children. It was back to Boulder again with only one new bit of information—the Treders had provided information that led to Jim Wright, who had moved to Denver and was a student of Colorado history. Perhaps he might know something. The Denver telephone directory listed several pages of Wrights, and twenty-seven James Wrights; an exhaustive search through the list of Wrights was decided upon only as a last resort.

Hours of poring over newspapers from Idaho Springs and Georgetown for news of McLean produced nothing of value. Georgetown's *Colorado Miner* did mention him occasionally in 1877. The stories were always that he had new photographs for the tourist of the period, or that he was departing on a trip to take more pictures. Further study of the negatives turned up one showing a picture of a Black child. Paul Stewart was interested in that and wanted to know how the McLean project was going. He knew a man named Jim Wright who might be able to help. He supplied Wright's phone number, and a call to Wright in Westminster, Colorado brought a surprise—he had a photograph of McLean himself, and it was signed in the photographer's own hand.

Now all the pieces fell together. George Treder did some fine detective work in this regard. He reasoned that in spite of the McLean spelling (rather than MacLean), the photographer was supposed to have been a Scotsman. In those days most English and Scots belonged to the Masonic Order. He contacted Merle Sowell, former mine superintendent and secretary of the Idaho Springs Masonic Lodge for 25 years, and a search of the lodge records uncovered the answers.

Before he came to the United States, McLean had joined the Masonic Fraternity by becoming a member of Union and Crown Lodge No. 307 A.F. & A.M. of Barrhead, Scotland, and affiliated with Idaho Springs Lodge No. 26 in 1903. According to Masonic records, McLean's first name was spelled Lachland, one of several variations that eventually appeared, and he was born in 1844 in Barrhead, Scotland [the correct date was actually October 3, 1843]. He decided to take up photography in Glasgow, Scotland, for probably the usual period of 3 years. On completion of his apprenticeship, he decided to go to the land of opportunity, America. The lodge records stated that he went to Colorado in 1872 and died May 29, 1929 [actually May 28] and was buried in Crown Hill Cemetery in Lakewood, Colorado.

There was still one item that needed to be nailed down. All of the Idaho Springs people insisted that he had lived in Central City, but no records had verified those reports. Old copies of the Central City *Daily Register* gave no clues, although a photographer listed in Central City in the early days did make stereoscopic pictures for the tourist trade using a McLean negative. Of course, that did not necessarily mean that McLean had lived there. He apparently sold his negatives and prints to anyone who wanted them and certainly was not zealous in claiming credit. One of his pictures even appeared in the 1876 edition of Frank Fossett's *Colorado, Its Gold and Silver Mines.* This was a year before he was listed as having a shop in Georgetown.

The question of whether he ever resided in Central City was finally answered when Jack Hidahl, public relations officer for Central City, checked through the old records and found that he was listed in the tax records for 1872 and had lived on the "Casey," the better part of town. The old-timers had been right. He had lived on "The Richest Square Mile on Earth," and his career had extended over half a century. Having been born in 1843, he must have been 28 or 29 when he moved to Central City. That could have been the start of his professional career.

So ended the quest for a pioneer photographer, one of many early Colorado chroniclers who helped to tell the story of the Colorado hardrock miner.

The District

Clear Creek Valley

. . . Only yesterday it seems the valley was a street of mines. A railroad with midget engines and dwarf cars whistled, rattled, and puffed hot cinders as its wheels screamed on sharp curves and its cars ground slowly up hill with cargoes of steel, dynamite, and food or downhill with loads of rusty or glistening ore. Six-horse wagons rumbled to the railway from side canyons carrying gold and silver hidden in mud and rock just lifted from underground. Tremors from subterranean blasting shook dishes on breakfast and dinner tables. Miners in stained overalls and high rubber boots plodded at daybreak to the mine portals and wearily homeward at dusk. The creeks were choken with yellow mud ground out of crushers in the ore mills.

Thus wrote James Grafton Rogers in the perceptive tribute to his beloved Clear Creek Valley, *My Rocky Mountain Valley*. The Clear Creek Valley was the place that Lachlan McLean spent most of his adult life, briefly on the north fork stretching into the hills beyond Central City, but mostly on the south fork that carves past the hot springs at Idaho Springs into the high mountain meadows beyond Georgetown and Silver Plume. For a short while he resided at Golden, a few miles east of where the two forks meet on their voyage to the sea.

This must have been one of the most beautiful valleys in Colorado before miners and settlers arrived. Clear Creek, deserving its name then, noisily leaped, wound, and washed its way down through mountain gorge and a valley deep with grass, presided over by quiet pines and rustling aspen. Breaking out of the mountains through a rocky canyon, Clear Creek eventually joined the South Platte River to roll across the prairie, far from its mountain birthplace.

Centuries ago Indians wandered into the valley and then, according to legend, came the Spanish miners searching for gold, a fact no known document records. The fur trapper found his way into the valley; a few may have wintered there, their passage scarcely noticed. They did not permanently mark the land, and the creek rushed on. Then came rumors of gold once more, in 1859, and the Clear Creek Valley and American history met. On the frozen banks of the stream two prospectors, insignificant mortals defying a winter in the mountains, found gold. These two dared and won. The valley would never be the same again. In an instant, time stood still, changed course, and the modern era was ushered silently into the valley. Man had come to stay.

Two men altered the saga of the Clear Creek Valley, southerners both, Missourian George Jackson and Georgian John Gregory. They prospected in these mountains because two groups of would-be miners had preceded them in the foothills in the summer of 1858, following will-o-the-wisp stories of gold, buttressed by a few facts. The 1858'ers found a little "color" beside Cherry Creek, where Denver one day would sprawl, and on their chance discovery was built the Pike's Peak gold rush of 1859.

If that had been all the gold, the rush would have been the hoax many claimed it was. Fortunately, for the reputation of future Colorado, these southerners and some others moved into the mountains to locate the source of the gold that had obviously washed down to the discovery site. They drifted in and out of the mountains during the course of the winter, while people by the thousands excitedly prepared to stampede from the states to Pike's Peak.

A veteran miner in his midtwenties, Jackson discovered gold in January, having dug down through ground thawed by a fire. The place became known as Chicago Creek and the diggings, the Jackson Diggings. James Grafton Rogers later called this the "most important single thing a man ever did in this valley." Tight-lipped Jackson kept his secret well, but when he and a group of friends returned in the spring, the rush was on. For the history of Colorado, the more important of the two men was another experienced miner, John Gregory, who in April uncovered a much richer deposit between future Black Hawk and Central City. He too left his name—Gregory Diggings, Gregory Gulch.

Neither of the men would loom large in Colorado mining after the first tumultuous months; they had played their role upon Colorado's stage. The flood tide of humanity their discoveries unleashed has ebbed and flowed over the Clear Creek Valley ever since.

Gregory Diggings proved to be the more significant, not only because they were richer in gold, but also because they were easier to reach. No wagon road, not even trails, ran into the mountains and certainly none dared challenge the narrow, rocky, and steep Clear Creek Canyon. To reach it one went over Gregory's trail, then took the cutoff. Among those who braved their way into Jackson's district was a farm couple, late of Kansas, Horace and Augusta Tabor. Augusta remembered:

> The road was a mere trail; every few rods we were obliged to stop and widen it. Many times we unloaded the wagon, and by pushing it helped the cattle up the hills. Going down hill was so much easier that it was often necessary to fasten a full grown pine tree to the back of the wagon for a brake. Often night overtook us where it was impossible to find a level place to spread a blanket. Under such circumstances we drove stakes on the ground, rolled a log against them, and lay with our feet against the log. Sometimes the hill was so steep that we slept almost upright. We were nearly three weeks cutting our way through Russell's Gulch into Payne's Bar . . .
>
> Ours was the first wagon through, and I was the first white woman there, if white I could be called, after camping out three months. The men cut logs and laid them up four feet high, then put the 7 × 9 tent on for a roof. Mr. Tabor went prospecting. I opened a "bakery," made bread and pies to sell, gave meals, and sold milk from the cows we had brought.

Tabor learned the rudiments of mining at Payne's Bar. He prospected and mined the summer away, leaving when warned that snowslides would endanger his family.

When he returned from depositing them safely in Denver, he found his claim had been jumped for want of the required monthly assessment work. Augusta lamented, "There was no law in those days." Actually, there was law, but it was mining district law and difficult to enforce in the transitory life of these early days. Tabor learned from this experience and eventually rose to epitomize the glamour of nineteenth century Colorado mining. Never again, though, did he return to invest in the Clear Creek Valley.

Only a short time passed before news of Jackson's and Gregory's discoveries rippled through crowds in the Denver area. Already an estimated 100,000 were heading for the "mines," probably half of whom quit before reaching Cherry Creek after hearing discouraging stories told by earlier arrivals, called "go-backers." Of the remainder, maybe half gave up after finding that the promised land was one of high costs, hard work, and a pan of rumors rather than gold. That left 25,000 or so to form the nucleus of future Colorado. Many of them rushed to Clear Creek.

Colorado's reputation was none too good at the moment, and the fortuitous arrival of that well-known New York newspaperman, Horace Greeley, helped save the day. Greeley eventually reached Gregory Diggings and sent a long account back east telling of the gold discoveries and painting a somewhat somber picture: "the hardest (though sometimes the quickest) way to obtain gold is to mine for it."

> As yet, the entire population of the valley—which cannot number less than four thousand, including five white women, seven squaws living with white men—sleep in tents, or under booths of pine boughs, cooking, and eating in the open air . . .
>
> And I feel certain that while some—perhaps many—will realize their dreams of wealth here, a far greater number will expend their scanty means, tax their powers of endurance, and then leave, soured, heartsick, spirit-broken.

Horace Greeley had "seen the elephant," as the miners loved to say, and correctly judged what the outcome would be for many of the participants.

From these beginnings grew the mining industry and settlement of the Clear Creek Valley. These men placer mined, searching for free gold in and along the stream bed. At first they used only a pan, a pick, and a shovel, augmented by much hard labor. Later, employing American enterprise, they switched to the rocker, long tom, and sluice to recover more gold for the same, or perhaps less, effort. In these procedures, they followed the footsteps of the 1849'ers in California who had faced similar problems a decade before. Scattered among these fifty-niners were forty-niners who knew what they were about and even some men who had

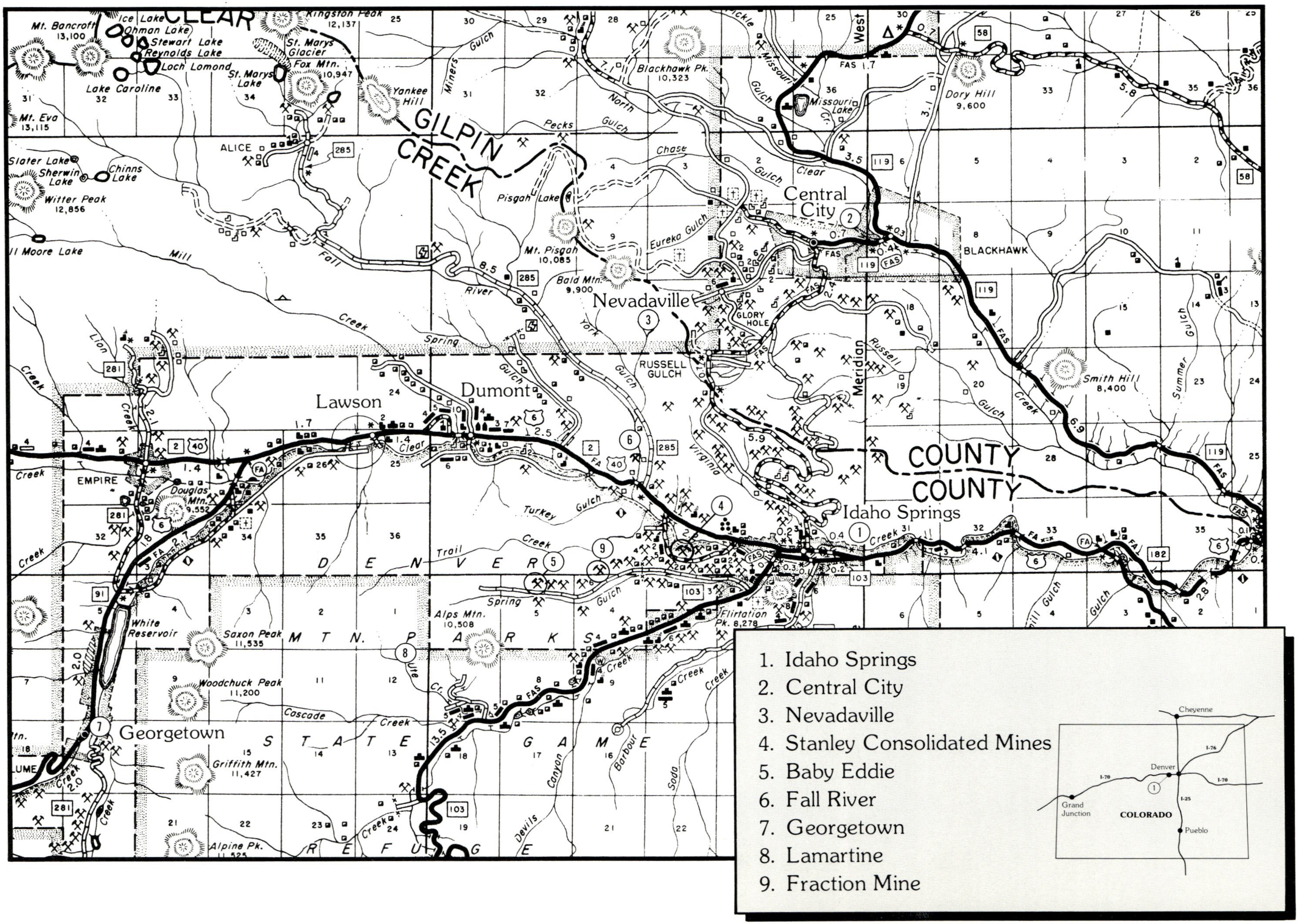
1. Idaho Springs
2. Central City
3. Nevadaville
4. Stanley Consolidated Mines
5. Baby Eddie
6. Fall River
7. Georgetown
8. Lamartine
9. Fraction Mine
Cheyenne
Denver
Grand Junction
Pueblo
COLORADO
GILPIN
CLEAR
CREEK
COUNTY
COUNTY
Central City
Nevadaville
Idaho Springs
Lawson
Dumont
Georgetown
EMPIRE
ALICE
BLACKHAWK
RUSSELL GULCH
GLORY HOLE
Mt. Bancroft 13,100
Mt. Eva 13,115
Kingston Peak 12,137
Fox Mtn. 10,947
Yankee Hill
Blackhawk Pk. 10,323
Dory Hill 9,600
Mt. Pisgah 10,085
Bald Mtn. 9,900
Smith Hill 8,400
Witter Peak 12,856
Douglas Mtn. 9,552
Saxon Peak 11,535
Alps Mtn. 10,508
Woodchuck Peak 11,200
Griffith Mtn. 11,427
Flirtation Pk. 8,278
Alpine Pk.
Ice Lake
Ohman Lake
Stewart Lake
Reynolds Lake
Loch Lomond
Lake Caroline
St. Marys Glacier
St. Marys Lake
Slater Lake
Sherwin Lake
Chinns Lake
Missouri Lake
Pisgah Lake
White Reservoir
D E N V E R
M T N. P A R K S
S T A T E G A M E
R E F U G E

mined in the Georgia gold fields back in the 1830s. The principle remained the same whether a pan or sluice was used: gold was heavy and if gently agitated would sink to the bottom, while the lighter sand and gravel washed away. The rocker, looking like a baby's cradle, allowed the miner to work more "pay dirt," and the long tom and sluice, similarly built wooden troughs, permitted larger amounts of dirt to be washed through, resulting, it was hoped, in more gold.

Frank Hall, who spent the season of 1861-62 at Spanish Bar, a mile west of Idaho Springs, wrote of his experiences. They were far more typical than those of the few who made fortunes and captured headlines.

> . . . a rough but helpful healthful and strengthening experience, full of knocks and bruises, of trials and disappointments, yet not wholly without profit, but in the six months of digging, wheeling, tunneling, and timbering, after consuming all the provisions and clothing we had brought from the States, we gathered not less than ten dollars worth of shining metal, with a harvest of knowledge that has lasted the better part of a lifetime.

Others made more money than Hall, some endured even more disappointment.

Placer mining went on for several years on Clear Creek, as Hall indicated, but the deposits did not prove as rich as in California. Soon, within weeks in the Gregory Diggings, miners were slashing away at hillsides and burrowing into the ground to follow gold veins deep in the mountains.

There were other problems that needed to be resolved as well. Clear Creek technically remained part of Kansas Territory, a government far removed. The miners turned logically to the extra-legal precedent of miners' law and miners' courts. For validating the ownership of claims, they established the mining districts. These were sufficient to provide the rudimentary, economical government that miners wanted, even if they left something to be desired, as Horace and Augusta Tabor found out.

The mining district boundaries were drawn as simply as geography allowed and of a size to make governance easy. Future Gilpin County comprised nineteen mining districts, for example. The initial governing laws reflected a similar wish for simplicity. The miners of Jackson Diggings held a meeting on May 9, 1859, establishing the size of a claim and stating that each claim should be marked and that the discoverer of a new digging was entitled to one extra claim. Others, apparently, could claim only one, by the right of staking it. This displayed a very real concern that greed would supersede common sense. The first party to find a new gold pocket might otherwise monopolize it and prevent others from sharing in his good fortune.

The miners at Gregory Diggings met on June 8 to write laws that proved more detailed, including the requirement that claims be worked within ten days of staking them. District boundaries were set up, along with the procedure for a miners' court to settle disputes. Nobody wanted to spend too much time away from the all-consuming pursuit of wealth. Yet these men, and the few women like Augusta, were doing more than simply mining; they were establishing settlements in Clear Creek and using it as a base to explore deeper into the mountains. Before the warm summer days turned chilly, they had gone as far as South Park and beyond into the Breckenridge area.

Many came just to mine during the summer season, others chose to stay. They retained faith that the gulches and bars (placer deposits in a slack portion of a stream) would be richer next year than this. Faith was a major ingredient of nineteenth century mining.

By fall, signs of this faith emerged everywhere. Embryo mining camps rose without plan near the spot where miners labored. The most promising and largest was Central City, which Greeley had visited. Around it grew Nevadaville, Mountain City (actually the oldest community), and Black Hawk. The mining frontier was an urban frontier. Other signs appeared also. In letters from Gregory Diggings in October, writers mentioned stamp or quartz mills, indicating that mining was becoming more of an industry, one that required investment. The ore being dug from the ground needed to be crushed before it could be worked for gold, hence the mills.

The need for law became more critical, as made evident by one who signed himself only as CW when he wrote to Denver's *Rocky Mountain News* (October 27, 1859) that a man had been convicted of stealing blankets and money. The miners' court sentenced the miscreant to thirty-nine lashes, the shaving of one-half of his head, and also "invited him to leave the mountains." Quick and economical justice was what the miners demanded.

In the following years the fervor abated in Clear Creek Valley. The early sixties were the last banner years for placer mining. Hardrock mining moved to the forefront, and with it came the foundation for corporation development and the end of the freewheeling individualism that marked those early days. Now it took more money to mine. The prospector and poorer miner moved on to find new diggings, and thus the cycle repeated itself. Empire, above Idaho Springs, was the last good placer area and it was finished by 1865.

New concerns confronted the settlers. In June 1859, a destructive fire rampaged out of control near Jackson Diggings, killing three men and bringing some caution to mountain living. This tragic lesson still had to be learned by other people in other places. Persistent transportation problems hampered the region for years, until the railroad eased them. Better trails replaced poor ones, then came wagon roads to Gregory Diggings. Not until 1861, however, was a road forced up Clear Creek to Idaho Springs and then over the mountain to Central City. It would be used for many years as the best and least mountainous of the roads.

As the Civil War neared in 1861, settlers worried about the "union" sentiments of their neighbors. Clear Creek stayed loyal, the southerners either departing or maintaining a quiet sympathy. Others doubted the veracity of miners; mining fever readily addled even the most otherwise scrupulously honest individuals. "G" wrote about the problem to the *Rocky Mountain News*, July 25, 1860: "It seems as if that after a man becomes a miner here [Gregory Diggings], he becomes at the same identical time constitutionally and conscientiously incapable of revealing the truth of mining matters or relating the straight facts." The region and state suffered mightily because of this syndrome.

All these things happened, of course, a decade before McLean arrived. In the mid-1870s, he would see only the remains, the ripped and scarred landscape. However, the heritage of these early days lingered and even this Scotsman would try his hand at panning.

Four communities sired by this rush remained and would play a role in McLean's life: Central City, Idaho Springs, Golden, and Georgetown. Of these, Golden was the least important to his story.

Augusta Tabor had camped near Golden for three weeks in July 1859, while her husband made his first prospecting venture into the mountains. Some mining was carried on in this area in 1859, but without the discovery of any lasting deposits. Editor William Byers of the *Rocky Mountain News* visited the fledgling community about the same time as the Tabors and was "astonished" to find David Wali's farm in such a flourishing condition. Wall, one of the first to make a success of agriculture in Colorado, located his farm ideally to tap both the mining camp trade and Denver. The town was likewise fortunate, but neither mining nor agriculture gave Golden its early fame.

Golden, the gateway to Clear Creek Valley, emerged as Denver's chief rival in the 1860s, and the two alternated as territorial capital for several years. Eventually it lost out to Denver's more aggressive and better financed businessmen and leaders. But not before it had attracted some industry, including ultimately a brewery, foundry, smelter, paper mill, flour mill, and fire-brick works. As the home base for the Colorado Central, Golden acquired railroad shops and transportation significance. Around the town developed coal mines to satisfy the growing need for fuel. In its own estimation, Golden was a prosperous manufacturing and business community by the time McLean saw it. The "Lowell" of Colorado, it was called by some because of its industry.

Beauty is in the beholder's eye and that belittling Englishwoman Isabella Bird saw little to praise during her 1873 visit. Golden, she wrote, has a "meanness" belying its name. Brick, pine, and log houses huddled together were not appealing to her, and she claimed that "every other house is a saloon." Hardly a woman was seen, a fact that annoyed this early feminist. She went on to decry a disgusting American habit: "Golden City rang with oaths and curses, especially at the depot. Americans are given over to the most atrocious swearing, and the blasphemous use of our Savior's name is peculiarly revolting." Golden survived Bird's tirade to play the smallest role in McLean's life of any of the Clear Creak communities.

The young photographer logically selected Central City first to establish his business. Until the emergence of Leadville in 1878, Central reigned as Colorado's largest and wealthiest mining town and the most significant of the mountain settlements. Central City had taken the lead from the start; even Greeley realized its overall importance as a mining district, and it maintained that position for nearly two decades. Not until Cripple Creek inundated the state and nation with its golden flood in the 1890s was Gilpin County surpassed as a Colorado gold producing district.

Central had grown in response to the needs of the miners along the narrow gulches of Gregory Diggings. No one plotted development, it just grew, a typical mining community. Sired by mining, Central City evolved from camp to town, living and dying with the fortunes of the industry. Writer and newspaper editor Ovando Hollister, who knew it firsthand in the 1860s, called Central "the metropolis of the mines."

Visitors evinced a mixed reaction to Central City. Active traveler Samuel Bowles, a leading American newspaperman, felt in August 1868 that, while it might be a good place to acquire gold, all beauty was sacrificed to use and no genuine homes existed there. He was wrong, many called it home. But he was right that mining was transitory. A decade later, an unidentified Michigan tourist thought the town presented a "new, substantial appearance," quite in contrast with other mining districts. The fact that Central had just rebuilt from a disastrous May 1874 fire helped. A correspondent of the *Engineering and Mining Journal* held a different view back in September 29, 1877, claiming the community ". . . enjoyed the distinction of being the most unattractive city to be found within a range of many hundred miles." He continued on in the same vein, commenting that the site "does not strike the visitor as either pleasant or advantageous." Clear Creek was contaminated and muddy, and the neighboring smelter town of Black Hawk polluted the air with "cough-compelling odors." Only the imposing Teller House, one of Colorado's finest hotels, warranted praise.

What was Central City? First of all it was a mining community, jerry-built of logs and wood primarily (until the fire of 1874) and squeezed into narrow gulches between barren hills. Miners had little time to worry about aesthetics in the rush to rip and tear the land in the scramble for gold. Nor did they consider the beauty of the tree-covered slopes, whose product was indifferently chopped for fuel, mine timbers, and lumber. Little thought was given to town planning; Central merely stretched where it could. As the center of mining, mining naturally dominated its life. Mining's desecration was tolerated for the sake of jobs and profits. Mining was a masculine dominated world, and Central answered the needs of that world for entertainment and business. The visitor might be shocked, but the resident understood.

The fairest way to try to recapture the community and its period is through the eyes of a person who lived there throughout the years McLean called it home. One-time cashier of the First National Bank, Frank Young left behind an absorbing personal recollection in *Echoes from Arcadia.* Let him give insights:

> . . . they [Central Citians] assumed to lead, and none even questioned their right. Their domicile became known after a time as the "Little Kingdom of Gilpin"—a title extorted from fellow Coloradans outside their borders, because of their industrial progressiveness, their social prominence, and their influence in public affairs.
>
> Our mountain retreat, we feel, is now no longer on—or indeed beyond—the frontier, as in the past [before the railroad, 1878]. We shall no doubt be overrun with tourists of all sorts and conditions—commercial drummers, pleasure-seekers, scientific bug-hunters, and possibly a capitalist now and then, with an eye to investment in gold mines . . .
>
> As a colony, independent of its surroundings, and sufficient unto itself—holding its own unique place and leading in all Territorial matters as though by inherent right—it has finished its career [1880, according to Young]. Its days are done; but its memory lives, and will ever be kept green in the hearts of the survivors of the cycle just completed. Its story is of exceptional interest . . .
>
> . . . you will find a Gilpin man or men in any mining camp, however distant or obscure, that you may chance to drop into.

Young expressed clearly the certain arrogance of prosperity, the confidence of youth, and the pride of success that were Central City of the McLean era. What outsiders might think or say seemed to matter little, unless by chance it adversely affected mining. In an offhand manner Young commented further on what he thought greatly influenced his contemporaries: "Surely there is nothing fantastic or strained in crediting the close environment of the mountains with a positive, moving influence upon human action."

Even with its wealth and prestige, Central played a lesser role in McLean's life than did rival Georgetown. While Central thrived on gold, Georgetown became Colorado's first "silver queen" (after a gold-generated beginning.) More isolated than Central, the Georgetown district developed more slowly and the camp languished, smaller in size. In 1864-65, when silver was discovered and mined, this district finally came into its own.

In 1867 Georgetown and neighbor Elizabethtown merged to become one town, keeping the former's name. Georgetown, in its own eyes, now stood as a competent rival to the better known Central City. Unlike its older competitor, Georgetown generally earned good reviews from visitors, probably a result of its more beautiful valley setting, nestled among mountains on three sides. The more open, spacious site allowed the town to grow in a more gracious manner, a fact not lost upon tourists.

Samuel Bowles called it one of the places every tourist should visit, partly because of its silver mines and partly because of its mountains and scenery, particularly that favorite of the nineteenth century, Gray's Peak. The 1870 *Rocky Mountain Directory and Colorado Gazeteer* gushed over the scenic grandeur, too, before going on to praise Georgetown. The streets were paved with gravel and "entirely free of mud" in all seasons. The climate received plaudits as unusually mild, considering the elevation, and "exceedingly pleasant and healthful." The author concluded by saying tourists would find Georgetown a "pleasant place to resort and mining operators and capitalists a good point for safe and profitable investments." What fault could possibly be found with it?

Visitor Isabella Bird found something. She was not impressed with this town of irregular streets crowded with miners standing in groups or drinking and gaming under "verandas." She rose to rhapsody about a trip to nearby Green Lake, however, and finally admitted that Georgetown was the only town in America which "deserved the epithet picturesque." From Bird that was more than faint praise. To balance her view was that of Frank Fossett in his *Colorado* (1879 edition). Fossett knew the state and mining districts better than Bird and could appreciate Georgetown for its worth. This was also the town that McLean knew and photographed.

> Its [Georgetown's] appearance and surroundings are superior to those of any other mountain town. Around it are lofty mountains, ribbed silver veins, which rise abruptly to heights of from twelve hundred to twenty-five hundred feet above the almost level valley in which the town is built. . . .
>
> The town possesses an excellent system of water works, an effective fire department, five churches, an opera house, and a grade school with some 360 scholars. There are two weekly newspapers, two banks, several hotels, and many business houses, some of which do a large trade. . . .
>
> There are many resorts near Georgetown that are unsurpassed in beauty and general attractions. . . . There are good hotels and fine livery turnouts in Georgetown, with pleasant rides and drives above and below town.

Lest it be surmised that Georgetown was primarily a tourist center, it should be kept in mind that it had been founded on the bedrock of mining and survived because of mining during these years. It exhibited some of the same attitudes and spirit as Central. Charles Harrington pointed this out in his *Summering in Colorado* (1874). He called Georgetown a "metropolitan mining camp," whose bustle and business suggested a grand system of mining operations. Like all mining centers, it survived alternate seasons of prosperity and depression, especially until the Colorado Central reached the town in August 1877. Miner, businessman, and photographer benefited as the train brought in investors, tourists, and suppliers more quickly and cheaply than the stage and freight wagon.

After his journey through the Clear Creek Valley communities, McLean finally settled in Idaho Springs in 1885, spending the remainder of his life there. Born as a mining community, Idaho or Idaho Bar, as it was called initially, was the first to turn unabashedly to tourism and market its scenery and hot springs. Even Isabella Bird liked it, the only community in the valley that she wholly approved of. (Although she did not take time to visit Central, it is not hard to imagine what her reaction would have been.) "A fashionable mountain resort," she called Idaho, as she rapidly passed through. On that off-season November day, it was practically deserted.

Starting as a placer mining settlement with a temporary future at best, Idaho Springs survived because of its tourist potential and an economy supplemented by the development of hardrock mining nearby. Of the three major Clear Creek Valley mountain communities, Idaho Springs was the least tied to mining and displayed the fewest mining traits. This was evident in many respects, a noticeable one being that Central City and Georgetown did not look upon it as a rival. In the day of nineteenth century personal journalism and local boosterism, the fact that a neighboring town praised another, as the *Georgetown Courier* did August 21, 1878, calling Idaho "a celebrated summer resort," indicated few hard feelings between the two. The reporter went on to describe the very prosperous conditions, " . . . guests finding plenty of amusement visiting the romantic, picturesque spots in the neighborhood; bathing at the justly celebrated mineral baths; riding, and out-door games." He even found time to praise local mining.

The mineral springs were Idaho's greatest claim to fame. Colorado was blessed with many but none better located to draw the Denver and tourist trade in a day when the railroad was supreme. Nineteenth century mineral springs were credited with health giving and health restoring capabilities that would raise medical hackles today. People flocked to them in expectation of near miracles.

Mineral springs, which he called soda, were observed by George Jackson during his prospecting venture in January 1859. According to another story, miners sinking a shaft in 1860 encountered hot water and were forced to abandon their workings. That abandoned prospect hole, used occasionally for bathing by locals and travelers, remained unimproved until 1863, when a small bath house was erected and Idaho Springs began its career as a health and pleasure resort.

What could these mineral springs, six in number, accomplish for the partaker of their healing power? Listen to this 1880 statement: "They are valuable waters—especially useful in rheumatism, cutaneous diseases, contraction of joints, etc." Traveler Samuel Bowles proclaimed that the wonderful hot soda springs furnished a most refreshing and health giving bath. The springs, it was said, were beneficial to invalids and healthful for everybody who took a plunge, dip, or swim. Testimony was gladly furnished by any local hotel or bath house to show the value of this mineral water. Come see, stay, and enjoy "days and weeks of unalloyed happiness . . . The devotee of pleasure, the admirer of nature, or the debilitated invalid may enjoy himself or find relief of the thousand ills which this weak flesh is heir to." When the Colorado Central reached Idaho Springs in June 1877, an even brighter day dawned.

Idaho Springs grew even more slowly than Georgetown. An 1866 visitor described it as a "straggling village of log huts." Until 1867 it had been the county seat, losing that important designation to larger and more progressive Georgetown, a setback survived because of development in the seventies. Although quick arguments would have been forthcoming from neighboring communities, it is probably safe to say that Idaho Springs' location was the best all-round of the Clear Creek mountain towns and its future prospects the least capricious.

Frank Hall, who mined there early in the sixties, wrote thirty years later that the wreckage of the mining frontier could still be seen from below Idaho Springs to Georgetown. The remains of log buildings, the long disused tramways, the ruined mills, and land "torn up and tunneled by hordes" of gold diggers signified that even Idaho Springs could not escape its past.

All these McLean saw, and he lived through the transition into the twentieth century. By the time of his death in May 1929, these communities had evolved from the vitality of the nineteenth century to the gentle, rural decay of the twentieth. Once the cornerstone, mining declined to a pale shadow of itself along Clear Creek, and interest in hot springs waned, submerged in the rush of new things and new interests.

During his fifty-six years, McLean experienced one of Colorado's exciting eras, as well as one of its transitional periods. Along the way he had resided in three significant types of Colorado towns: Central City (mining), Georgetown (mining, some tourism), Golden (business, industry), and Idaho Springs (tourism and mining nearly equal). Georgetown and Central were much alike, while Golden and Idaho Springs offered very different economic bases. The major mountain communities all declined noticeably in population by the 1920s, a development they lamented. Regardless, fate proved kinder to them than to many of the smaller camps McLean had visited on long-ago photographic excursions; they no longer existed.

These urban beacons of the mining frontier had helped to transform the Clear Creek Valley irrevocably. Before them, no settlement had stirred the land. Now the wilderness and frontier were gone and people settled to stay. Georgetown, Central City, Idaho Springs, and, to a degree, Golden brought this about. They had nurtured the rudiments of civilization, tending them until they flourished as opera houses, schools, literary societies, newspapers, businesses, churches, and a host of other ingredients. They promoted and encouraged regional advertisement, settlement, industry, and tourism. To tap their potential market and nearby natural beauty, the railroad came and with it great change. As Frank Young wrote of its smoky debut in Central, "We act much like children on the advent of the railroad." He only briefly mourned the passing of the stagecoach and the epoch that ended with it.

Two generations of Coloradans lived in these towns during McLean's photography career. He know many of them, their towns, and their times. His historic role in all this was one of preserving through his photographs an image and a feeling before they passed from the scene. Self-effacing, McLean did not intrude upon his subjects and hardly can be traced from the places he lived, yet he left a heritage for everyone who takes the time to examine his work. A light in the dark night of history, his photographs illuminate the silent past, making it vivid once more, making it a part of our today and everyone's tomorrow.

Population Statistics

	1860	1870	1880	1890	1900	1910	1920
Golden	1014	587	2730	2383	2152	2477	2484
Central City	598	2360	2676	2480	3114	1782	552
Georgetown	---	802	3294	1927	1418	950	703
Idaho Springs	---	229	733	1338	2502	2154	1192
Clear Creek County	---	1596	7823	7184	7082	5001	2891
Gilpin County	---	5490	6489	5867	6690	4131	1364

It would appear that the population reported for Central City in 1860 was low and that for Golden high. The logical explanation is that many miners chose to winter near milder Golden or Denver or back home in the states and had not returned to the mountains when the census taker appeared. Georgetown's peak total in 1880 and subsequent drop reflect problems encountered by silver mining. Central City gained people in the late nineteenth century, as the smaller camps around it declined and in some cases died. All the mountain towns and counties lost significant percentages of their population in the decade from 1910-1920. By 1920 Clear Creek and Gilpin counties ranked 51 and 59, respectively, among the 63 Colorado counties in population.

The Man

Lachlan McLean

The search for the elusive Scotsman Lachlan McLean would have done justice to Sherlock Holmes. This apparently retiring photographer left few historical tracks to follow, as the Prologue amply describes. Even the year of his birth and the spelling of his first name (Lachlean, Lacklan, Lachlan, or Lachland) remain in question. The original 1900 census listed October 1843, a date supported by his obituary in the *Clear Creek Mining Journal-Gazette* of May 31, 1929, which gives it as October 3, 1843 [so does the death certificate]. However, Masonic records and the Colorado census of 1885 say 1844. To confuse the issue further, the 1880 U.S. Census dated his birth in 1847, which puts in doubt the ability of the various census takers or the answers provided by the persons interviewed!

Confusion over the spelling of his first name brings into conflict the Masonic records (Lachland), the 1880 and 1900 census (Lachlean), the obituary (Lacklan), and the certificate of death (Lachlan). If this reticent Scotsman had intended to give future historians gray hairs, he could not have succeeded more admirably.

As a starting point, the obituary placed his birthplace in Glasgow, Scotland (Barrhead is just a stone's throw to the southwest), where he grew to young manhood. According to this account, he came to America in 1866, settling in Denver, presumably to work as a photographer, because the article mentioned that he took his first picture of Idaho Springs in 1869. McLean was employed for a time by railroad companies to take views along their lines, not an unusual practice, although it is unclear just when this activity occurred. In 1871 he went home to Scotland to join the Masonic Order and shortly afterward returned to America, settling in Pennsylvania before traveling back to Colorado and Central City. The source of this information, probably one or more of his children who heard it after the fact, stated that he married Anna Ellson in 1873, then came west.

Colorado evidence for any of this is lacking. A careful examination of Denver's 1870 census returns and business directories did not disclose any L. McLean, as he was usually listed, nor did a sampling of the *Rocky Mountain News*, 1866-69, although it did carry advertisements of several photographers and mentioned others in news items.The absence of information in these sources is not unusual and does not necessarily discredit the obituary. The young immigrant may have worked for another photographer to learn the business, and in 1870 he may have been out of town on a photographic expedition. Perhaps he was simply missed by the census takers, a common occurrence.

The census of 1900, the most informative on the McLean family because of the questions asked, gave 1871 as the year Lachlan immigrated to the United States, where he later became a naturalized citizen. No mention was made of any earlier visit. His wife Anna came from England in 1872, which gives validity to the

1873 marriage date, especially since the same census listed their being married 26 years. Where they met and married remains unknown. It is conceivable that McLean came to Colorado first, then went back to Pennsylvania to be married.

At this late date it is a tempest in a teapot; the marriage seems to have been a happy one. Six children were born to Lachlan and Anna—Robert, Willie, Minnie, Norman, Harold, and Mabel—dating between 1875-1890. Anna died just before her husband, a fact seen to have hastened his death, along with grief over the recent death of one of his sons and an injury McLean himself incurred sometime in the winter of 1928-29.

Up to this point no historic moorings have been found, but in the 1870s McLean finally comes into view. Let not the reader think that he appears clearly—the historic shadows of a century continue to blur any vivid outlines.

He paid taxes in Gilpin County in 1872, listed himself in an 1874 Denver business directory as "L. McLean, photographer," and in 1876 had a studio opposite the Teller House in Central City. A year later he appeared in Georgetown, and following a one-year sojourn in Golden in 1884, settled in Idaho Springs for the remainder of his life. This is the bare outline of the life of one whom the *Journal-Gazette* called an "old citizen."

Just how long McLean may have been in Central and the duration of his Denver stay are unknown. Newspapers from each place provide no help. Regardless of the community, he found professional rivals already on the scene in each one. Fellow Scotsman Joseph Collier in Central was one of Colorado's best photographers and eventually came to monopolize the business in Gilpin County, until he sold out and moved to Denver in 1878. Perhaps Collier's competition proved too strong for the younger man, explaining the move from Central. Until the arrival of the railroad and steadily increasing numbers of tourists, photography offered only a narrow and slowly expanding market. Although there was not a large number of photographers in Colorado in the 1860s and 1870s, competition belied the total.

Adventurous photographers appeared almost in the wagon dust of the first fifty-niners. Newspaperman William Byers saw one at work in the Denver area in June 1859, and another opened a business in Central City the next year. William Chamberlain located in Denver the same year, 1860, and for the next two decades prospered, being especially noted in the 1870s for his mountain views. McLean must have known him, certainly his work, and he, too, would take many scenic photographs to sell to tourists. Obviously Lachlan was partial to views along the Clear Creek Valley railroad lines. The ones which remain, of those he took, were typical for his era.

The best known and most famous of McLean's contemporaries was William Henry Jackson. He first photographed Colorado in the 1870s for the Hayden Survey at a top salary of $175 per month. Moving to Denver in 1879, he entered into private practice with a yearly income, by the late 1880s, of about $9,000. This gives some idea of what McLean must have earned, for even at this date Jackson was well known and living in the state's largest city. In smaller, much less thriving Georgetown and Idaho Springs, Lachlan could count on probably a third of Jackson's income. Not a poverty wage by the standards of the day, it was certainly far short of what men like Tabor generated from their mines, and many merchants from their stores.

Jackson has left behind a vivid account of the tribulations of 1870s photography during the days of the "wet process," when the glass plate was photosensitive only as long as chemicals on it remained damp. Each photographer had to be his own chemist, preparing his own plates. A student of Colorado photography, Terry Mangan, in his *Colorado on Glass*, observed that, considering the photographer had to haul tremendous numbers of glass plates, pitch a tent, prepare the plate, set up the tripod and camera, return to the tent and develop the plate, it was a wonder any photographs were taken away from the gallery. Jackson, who just might have been the best at field work, described it:

> . . . how onerous were the conditions imposed upon the old-time photographer. Then, the so-called "wet plate" was in general use, involving a complex process of many operations and requiring practice and experience to attain the necessary manipulative skill. It was so messy chemically, and encumbered with so much apparatus that there is little wonder that photography was rarely the hobby of amateurs and that the making of pictures by this means was almost wholly the work of professionals. . . .
>
> The preliminary planning to provide well balanced supplies was a matter of some care, as, once afield, there was slight possibility of making good any deficiency. . . . The glass alone—some 400 pieces, was no inconsiderable item, while the collodions, silver nitrates, iron sulphates, and a score or two of other chemicals and articles, made up in bulk what they lacked in weight. . . .
>
> With everything ready, I took one of the plates inside the tent and proceeded to flow it with collodion, which when "set" was immersed in the silver bath until the proper chemical reaction had taken place, when it was placed in the plate-holder ready for exposure—an operation that usually took about five minutes. . . .
>
> With exposure made and back to the tent, the plate was flooded with the developer by a dexterous sweep, the image appearing almost immediately. It was then well rinsed and "fixed" with the cyanide solution. . . .

When "hard pressed," Jackson could do it all, from unpacking to repacking, in fifteen minutes; a longer time was much more typical. McLean followed the same routine in the 1870s, until the day of the dry plate, and photographed some of the same scenes as Jackson. Unfortunately for him, he never became as famous.

Jackson appreciated what his contemporaries had

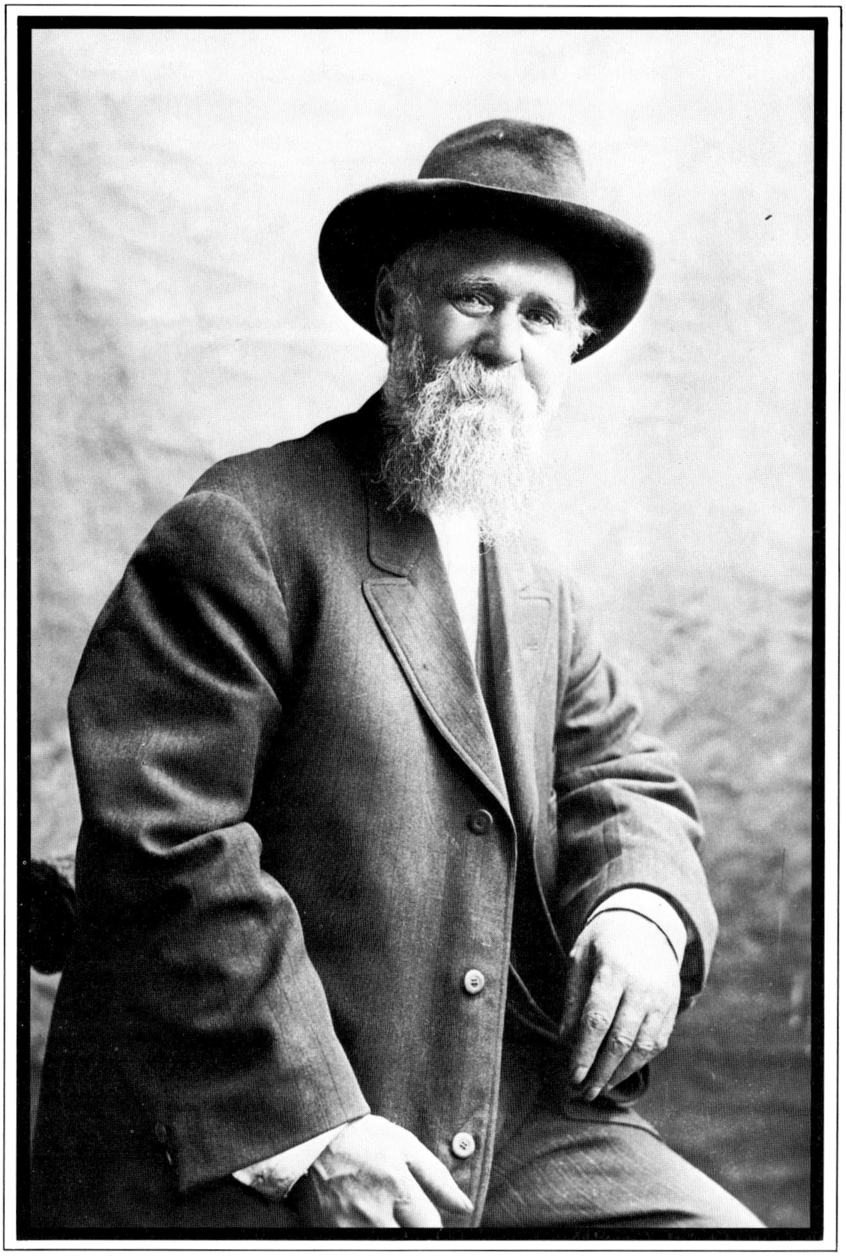

Lachlan McLean as he appeared in later life. It was said of him that he was a man of "sterling character," and "his kindly greeting" was a lasting memory to the people of Idaho Springs.

accomplished and tipped his hat to them in a 1926 article:

> In closing I want to contribute my bit of appreciation of the work of those other pioneer photographers of the early seventies, who in their day penetrated many of the then remote places in the Colorado Rockies and brought back excellent reproductions of its beauties and wonders for local demand.

Mountain panoramas took the photographer into the out-of-doors, which Jackson loved, in the summer and fall, and these scenes sold well to tourist and resident alike. Another hardy Denver photographer, Alexander Martin, explained how the times changed and the popularity of mountain scenery shifted:

> We photographers did a wonderful business in those days selling views, which brought prices far exceeding . . . prices of today [1919]. Then the newspapers began to install machinery by which photographs could be reproduced successfully by the halftone process, and the business began to decline. When the picture postcard . . . became popular, the view business, so far as the photographer was concerned, became almost of no value. Most of the photographers then began to limit themselves to portrait work.

Martin should also have mentioned, perhaps, that the popularity of the Kodak camera and its easy use for the amateur had something to do with this transition. McLean was good at portrait work, according to reports, although few of his efforts are known to have survived. Jackson considered portrait work the "bread and butter of every photographer." He was probably right, although for his own pleasure he enjoyed being in the field.

McLean's work at Georgetown corresponded with that of his Denver contemporaries. In August 1878, for example, the *Colorado Miner* reported that he left on a trip for Gray's Peak and the Mount of the Holy Cross. The public, the editor concluded, "may expect a treat in looking over his selection" on his return. Three weeks later, views from this trip were available at McLean's gallery on Rose Street. In a rare burst of publicity, the *Miner* carried four McLean items that month, including places where his "celebrated views" of Colorado might be purchased. For Lachlan personally, the proudest moment was the following item in the issue of August 24:

> Born
>
> In Georgetown, Friday, August 23 to the wife of L. McLean, a son. [Robert] We congratulate McLean on this addition to his family.

McLean was seldom mentioned in the newspaper. He only rarely advertised or printed a business card and apparently was not a joiner, except for the Masonic order. Neither he nor his family demonstrated a newsworthiness on the local social circuit, nor were they particularly friendly with the newspaper editor. For instance, the birth announcement that appeared just above Robert's contained four verbose lines; however, McLean must have had some standing, since the one after simply listed the birth, without comment. Lachlan seems not to have taken much interest in mining speculation and none whatsoever in politics, if the lack of newspaper commentary serves as a trustworthy guide. Nor was he involved in lawsuits, sports, or any public church activity. The McLeans, father, mother and children, were a very private family.

Occasionally, the *Miner* used a drawing of one of McLean's photographs, as it did for the front page of the August 3, 1878, issue. Georgetown from the south was the subject. As mentioned earlier, one drawing identifiable to a McLean photo appeared in Frank Fossett's publication. "Identifiable" is the key word, because others may have been used which now cannot be proven to have been taken by Lachlan. If he took photographs of something unusual, it might make the newspaper. The *Georgetown Courier*, December 19, 1878, described how "the frost king played a magnificent trick" on one of the large plate glass windows of the Clear Creek County Bank, the frost taking the shape of an "old gnarled, broken tree." On the spot, McLean photographed it and made stereoscopic views for those who had missed nature's handiwork.

Hardly a middle class Victorian home did not have a stereoscope in the parlor with a large number of stereographic cards nearby. The three-dimensional effect obtained when the viewer looked at a card with its two photographs of the same scene, taken from slightly different angles, was a wonder of the age. Popular from the 1870s into the 1890s, stereographic cards have since become a collector's item. McLean advertised "Stereoscopic and Landscape views, a specialty."

By 1880 McLean was well established in Georgetown, living in a predominantly middle class neighborhood on Rose Street. Among his neighbors were two blacksmiths, a carpenter, mining agent, and boot-and-shoe manufacturer, all of whom were in the twenty-to-fifty years of age range. The McLeans, and their three Colorado-born children, had lived longer in this community than in any other during their married life.

The January 1, 1881, *Colorado Miner* carried a McLean advertisement, part of which was quoted above. His gallery, like his home, was still on Rose Street, and he promised to give "prompt and careful attention to orders for mining and scenic views." He continued making trips into the mountains when the weather permitted, as it did, for instance, the following October when he set off for the Peru district, south and west of Georgetown. Sometime in the late spring, McLean gave up his independent gallery and went in with Robert Morrison. Morrison's Photograph Gallery, the establishment was called, with L. McLean listed as photographer. Robert Morrison, who worked in Georgetown in 1881-82, was much more aggressive, if his advertising methods are any evidence. All types of photographs, from cards to life-sized; all kinds of pictures taken and enlarged; and outside views of mountain scenery and mines taken to order were promised, with satisfaction guaranteed. Without blushing, he stated that his gallery had the "finest set of instruments in the State." To increase business, Morrison served as agent for the New York and Denver Portrait Company, which "guarantees the only durable crayon, ink or water color portraits made in Colorado." Why McLean did not try similar promotion will never be known. Perhaps his professional reluctance or personal aversion to such activities prevented it.

For all their efforts, Morrison and McLean did not remain in business very long and by December 1883, the previously quoted Alexander Martin had a gallery on Rose Street, probably in their old establishment. Martin had worked prior to this in Boulder and Central City, showing the same gypsy tendencies as McLean and others of their profession. Perhaps these moves were dictated by the search for new challenges, individual wanderlust, or economic necessity. A strong case can be built for the last argument in nineteenth-century Colorado.

A short stay in Golden, after leaving Georgetown, did not furnish the peace of mind or economic security Lachlan sought; thus a Colorado census taker found him in Idaho Springs in 1885. Again he settled in a middle class neighborhood of merchants, surveyors, druggists, and real estate agents in the thirties' and forties' age bracket. His family had now increased to four children. Their life there remains just as shadowy as that of previous years. A member of the family became an occasional subject of a news item, as did unfortunate Willie, who fell over a fence and broke a leg in two places while playing in a game of fox and hounds. Harold and Mabel were born in Idaho Springs, the last of the six children.

Lachlan held to his policy of not advertising, even when a Denver photographer periodically placed an ad in the local newspaper. His name appeared rarely, usually associated with a trip he was about to embark upon. Many of his photographs, however, were used by the Chamber of Commerce in 1896 and 1902 pamphlets praising the glories of Idaho Springs.

Varying the pattern, McLean did not move after a few years. He stayed and the family grew up there. His choice was wise—Idaho Springs' future was better than Central's and Georgetown's at the moment. As late as 1900, he rented a home in his usual middle class neighborhood, which was a mixed area of renters and homeowners. By that year the oldest two children had left home, and two others had taken jobs, Robert as a miner and Norman a painter; the McLeans could be proud of their family. All six of their children were still living, somewhat unusual in those days before modern medical practices and medicine greatly reduced the number of childhood deaths.

McLean maintained a studio in Idaho Springs for another twenty-seven years, gently sliding into old age along with his community. He and Anna had witnessed the frontier years; now they experienced the twentieth century, quite a change of pace for both of them. They basked in the title of pioneers, even though they were not members of that nearly vanished breed, the fifty-niners.

It was during his Idaho Springs years that McLean made his greatest historical contribution, his photographic record of mining in the Clear Creek Valley, centered around his hometown. He had been taking photographs of buildings and miners for years, but now with the improvement of photographic equipment, he went underground to find them at work. Herein lies his greatest significance, because nineteenth century underground mining pictures are rare. Obviously some of his shots were posed—men in suits and ties did not work a shift in the mines, even if they held some equipment or examined some ore. The sweaty, dirty miner did the work.

Most of the photographs are undated, although because of the general prevalence of candles for light, they would seem to be of the late nineteenth or early twentieth centuries. They are also unidentified as to place and people involved, yet this does not lessen their overall importance. They stand as a tribute to the men and industry which had opened and settled the Clear Creek Valley in the past generation. The photographs are also universal in their appeal. What McLean saw on his excursions was also taking place in Arizona, Montana, Nevada, and elsewhere throughout the mining West.

His obituary called Lachlan McLean an "expert photographer." He well might have been. His remaining photographs show a better-than-average ability, with a clear sense of composition. Where he really rose to the "mountain top" of photography, though, was in the underground mining photographs. The problems of transporting equipment to the site and lighting make McLean's mastery of the art even more impressive. In a very early issue of the *Rocky Mountain News* (May 15, 1861), editor Byers admonished his readers to have their pictures taken: "secure the shadow, ere the substance perish." That serves as a fitting tribute to the work of Lachlan McLean.

The Industry

Gold and Silver Mining

For the first fifty years of their existence, Clear Creek and Gilpin were two of Colorado's major mining counties. As Frank Fossett wrote in 1879, they were "in reality one large and continuous mining district." The placer mining deposits of 1859 and a few years afterward were quickly panned out. Although panning and sluicing went on for decades thereafter, hardrock, or underground, mining became the backbone of the industry. Mining developments in these two counties paralleled those elsewhere in Colorado and to a degree those throughout the whole of the mining West. Clear Creek and Gilpin counties both had their moment on center stage, then ebbed and flowed against a gently declining curve, until finally relinquishing their roles as stars of a mineral bonanza.

Befitting its circumstances in 1859-60, Gilpin County gained and held the prosperity lead over its neighbor throughout the 1860s. The switch to hardrock mining, noticeable by the end of 1859 with the introduction of stamp mills and the first attempts at following veins into the hillsides, came easily. At first it seemed nearly as simple as panning, if slightly more expensive and requiring more skill. The picks chipped away and stamp mills hammered to crush the ore. Dreams of quick and easy profits ended quickly and with them the first flush years for Central and its surrounding camps. As the miners dug, they first encountered the secondarily enriched ore near the surface where water and weather had leached out many impurities, leaving easy-to-work gold. Then came the awakening. As they went deeper, they sooner or later encountered what was described as refractory ore, which could not be profitably milled with methods at hand. High assay reports notwithstanding, the mills proved unable to save the ore, or could not work it at a profit for both miner and millman. Other miners encountered cap rock, either barren or low grade, and lost the vein they were following. By 1863 the bonanza times were gone, and money to experiment with new milling techniques or deeper mining was hard to find.

Deep mining was expensive. Deeper shafts and tunnels encountered water, necessitating drainage or pumping, and, at the same time, requiring more experienced miners and extensive machinery, neither of which Colorado possessed in abundance. As nearby forests were decimated by lumbering, the costs of fuel and mine timbers rose. An old story not unique to Gilpin County, it seriously hampered local mining when compounded by milling problems.

At this point, easterners rode to the rescue, or so it seemed. The Civil War, which had been raging "back in the states" (as Colorado papers referred to them), created a booming economy in the North, one with investors looking for a safe venture in a time of inflation and uncertainty. Nothing seemed more secure than gold. To eastern speculators a mine was a mine and if it contained gold, it represented a reliable investment that would turn a profit. No matter that a mine might be a shallow prospect hole and the gold more in the imagination than in the hand, the easterner wanted to buy. Mine owners obliged him. Gilpin County went through a brief flurry of mine and stock speculation in late 1863 and early 1864.

When the bubble finally burst in April, the aftereffects lingered for years. The *Engineering and Mining Journal,* August 14, 1875, commented that the county was not entirely free of the effects of the "disaster incurred during the old flush times." Even a decade later, concluded the author, many eastern men hesitated to invest or put more money into properties that had been only failures or worked at a loss. Many of the lodes were too small in size to be worked alone, and their eastern owners had been burned too many times to try again. The writer saw a golden cloud in the storm, though; consolidation was the answer. Then Central, ". . . as a gold producing district, will have no equal in the West."

The midsixties proved to be hard times for Gilpin County, partly for reasons of its own making, but also because of eastern expectations and demand for instant profits in those investments. Companies sent out managers who hardly knew a gopher hole from a mine shaft. Their idea of mining was to build a fancy mill without testing the process on their ores and to plan elaborate mine operations without adequately sampling ore reserves. Some of these activities represented fraudulent practices, many of them only carelessness and inexperience. No matter, eastern investors blamed Colorado for the failures, and interest in mining investment died.

The major problems still were not resolved. Following the collapse of eastern investment, it was hard to attract interest or money to help the hard-pressed Coloradans. Many of the mines were simply shut down when instant dividends were not forthcoming on the first stagecoach east. Perceptive Frank Fossett capably summarized what had transpired:

> Meantime very many mining companies had taken steps to work their properties, or at least get rid of their working capital as speedily or foolishly as possible. The entire history of these company investments and operations, with a few exceptions, could hardly have been worse. . . .It took years of depression and abandonment before the time arrived when these false steps could be remedied by consolidation. Poor management, foolish expenditures, or a failure to work the mines, with high prices of labor and supplies, were the other main causes of the failures of the companies. Of course, there were some claims that contained nothing of value; but this was not the case with very many of them.

Fossett was right; the *Engineering and Mining Journal* was right—small consolation to the miners in Gilpin County.

The situation began to change when Brown University chemistry professor Nathaniel Hill was sent by a group of capitalists to investigate their recently purchased claims. Hill became intrigued by the problems confronting milling, and out of his interest emerged the Boston and Colorado Smelting Company in 1867. A year later the firm opened a smelter at Black Hawk. The incompetence, the waste, the wild experimentation that had characterized previous attempts were replaced by the studious Hill's careful experimentation and testing. Over the next few years, he developed one of the outstanding smelter plants in the West. Of the new developments Frank Young said, "Many costly and fantastic constructions have gone to swell the junk-heaps of the foundries. . . The work of clearing up the debris and making a fresh start is necessarily slow . . ."

The fresh start was now underway, and Hill held the beacon in the night. Mining methods and equipment improved with the passage of time, as did miners' skills with the immigration of large numbers of those most capable Cornishmen, nicknamed Cousin Jacks. Central had turned the corner. The arrival of the Colorado Central Railroad to Black Hawk in December 1872 marked another important milestone; that it took six more years to reach Central was more a reflection of the national depression of those years than it was a lack of potential business. Successful mining required three ingredients, of which Gilpin County now had the first two: deposits of valuable ore, cheap, year-round transportation, and money to finance mine and smelter development.

The importance of Hill's smelter and the railroad to the revival of mining in the decade of the seventies can never be underestimated. The county produced an average of over $2 million worth of gold per year, and, as soon as the growth started, investors and outside money returned. Although Central was no longer the major mining district in Colorado by the end of the decade, it held its own very well, particularly for a district past its twentieth birthday. Mines are never notoriously long-lived.

McLean lived in Central during its most prosperous and exciting period since the opening rush. Mining was all around him, a fitting baptism for the Scottish photographer.

As Georgetown had been a step behind Central in opening its gold deposits, so it was several steps behind in emerging as a major mining district and town. Not gold, but silver turned the trick. Silver was mentioned as early as 1859-1860, but it was not until after discoveries in 1864 near Georgetown, and across the range in what became known as the Snake River mines in Summit County, that a silver rush unfolded in 1865. The new mineral came at an opportune time—gold production lagged, mining slumped, Colorado's reputation suffered following the gold speculation fiasco, and the territory badly needed an economic transfusion. Silver appeared to be the necessary elixir.

Circumstances dictated otherwise. A grand rush to stake claims ensued, as promises and expectations soared as high as the nearby mountain tops. After mining through a very shallow level of secondarily enriched ore, the miners dug into silver ore that defied all attempts to refine it. By 1867 the promises of two years before had stagnated. Mine development outstripped reduction advances, isolation slowed everything, needed outside capital did not come or was wasted, and Colorado's first "silver queen" looked a little the worse for wear, even though she was still in her youth.

Resolution of the ore smelting riddle held the answer. Silver proved harder to refine than gold, a fact already amply demonstrated on the much richer and better known Comstock Lode in Nevada. Stamp mills had been available in Georgetown since 1861; something more was needed. Sending ore to Black Hawk was costly, less so than to Europe, which some miners did with their high grade ore. In the end, they attempted to resolve the problem at home, a costly and time-consuming experiment. The results were put into words in the July 28, 1869, issue of the *Colorado Miner:*

> These "compliments to the dead" will not be found to resemble marble, nor nothing grand and imperishable, but rather will they be found after their pattern of ruined and *ruinous* mills, surrounded by rusty machinery, and decked with scattered fortunes.

Eventually, Nathaniel Hill's experiments solved the silver smelting problems, but not until the 1870s, after nearly a decade of frustration and expense.

In desperation, Georgetown miners turned to tunnels in an attempt to lower the cost of mining, hoping to produce the needed profit margin. It seemed so much easier and cheaper to drive a tunnel into the side of a mountain and intersect numerous veins than to sink a shaft down a vein, hoist the ore, and pump out the water. The resulting "tunnel craze" of the late sixties and early seventies proved to be more costly than expected. It was susceptible to speculation, because easterners failed to comprehend that a tunnel company did not own all the veins encountered. Although many companies were organized and filed on tunnel sites, most never got started; a few collected money and did nothing. Georgetown suffered.

Georgetown needed more than just a passing fad to place its mining industry on a solid foundation. The pattern and needs were obviously similar to Gilpin County's. Isolation (until the railroad came) and high freighting costs raised the price of everything and thwarted mining more here than at Central. Exaggerated expectations, poor management, and wasted money and efforts frustrated both districts. Inexperience with silver mining slowed development a bit, but not fatally, nor did the fact that the public seemed more interested in Colorado's gold than in its silver, until Leadville. Georgetown simply had to wait until the key factors of transportation and smelting were resolved. Waiting was, and is, one of the hardest things for a mining district and community to do.

In the 1870s all the pieces finally fit together to allow Georgetown a solid claim as the "silver queen." The greatest silver production in the sixties, an estimated $500,000 in 1869, soared to over $2 million for several years in the seventies, in spite of a declining international silver price.

Even in this moment of final triumph—"flush times," with mines producing, capital flowing in, and smelters successfully refining—Georgetown's joy was tempered. Rich, or potentially rich, mines have a way of attracting lawsuits; two of the richest were the Pelican and the Dives, above the neighboring town of Silver Plume. Frank Fossett recorded the struggle as a war, the almost "endless litigation between the Dives and Pelican."

> As the yield of both mines grew larger and larger the conflict became something like a battle of giants, in which services of a small army of fighting men, leasers, and guards were secured, along with nearly all the leading lawyers of the State, and some from the Pacific Slope. Suits and counter suits, injunctions and counter injunctions followed each other in rapid succession, and much of the enormous revenues that came out of the mines went to the lawyers or were divided among supernumeraries and lessees.

The four-year dispute reached its climax when a part owner of the Pelican was killed by an employee of the Dives in Georgetown. This tragedy eventually led to the end of the struggle.

Almost all the prominent mines in the Georgetown vicinity were, in one way or another, involved in lawsuits. Though not as bitter or as long as the celebrated Pelican vs. Dives case, they proved equally costly in time, money, and district reputation. Fortunately, the second half of the 1870s saw consolidation of most of the prominent mines; a twenty-five year period (1872-97) of steady silver production resulted. Overlapping claims, lost claim markers, poor records, and failure to record claims—all a heritage of the hectic sixties—levied a costly penalty in the next decade, abetted by an ignorance of geology, a subject in its infancy.

By the time McLean arrived in Georgetown, more caution was being exercised in mining and smelting; consequently, there was less boom-and-bust excitement and more profits and a more stable economy. He resided there, however, when another factor put a damper on Georgetown's exuberance. Uneasy rests the crown of a "silver queen," and usurpers came forth. Caribou, in Boulder County, proved easy enough to put in its place, but not so Leadville. The excitement of 1877 became the rush of 1878 and then the flood of 1879, pushing Leadville as a rival to Nevada's Comstock and Virginia City in fame and production, something Georgetown could never claim without tongue-in-cheek. Georgetown's reign ended abruptly; she became a dowager at best.

McLean basked in the last days of fame when it might still be said, as the *Georgetown Courier* did on February 26, 1880, "Clear Creek has over treble the number of paying mines that Lake County has. They do not pay as large dividends to be sure, but they will be paying when Leadville deposits have become a matter of history." It was a little late in 1880 to claim that much. Local pride died hard.

Georgetown was hurt by Leadville's brilliance. Money and publicity, which might otherwise have gone to the older district, now went to Leadville, along with eastern investors, who flocked to the flashier district like moths to a flame. Production and dividends, so dear to investors' hearts, could not compare to Lake County's.

Georgetown's day as a major Colorado mining town was over almost before the idea was accepted. It continued as the major Clear Creek County mining community and one of Colorado's consistent silver districts until past the turn of the century. McLean lived in the years of transformation when it slid from the flush of youth to the respectability of middle age. Potential and boom growth evaporated, for miner, photographer, and merchant alike. Unlike some of its neighbors, Georgetown held off the inevitable mining decline for several decades. Perhaps McLean sensed its certainty, which hastened his decision to move.

McLean spent most of his years (1885-1929) in Idaho Springs, but not because it would eventually emerge as the greatest of the Clear Creek mining communities. Actually, it trailed far behind his two previous homes in mining significance. When last mentioned in regards to mining (1865), Idaho Springs' thriving days as a placer district were behind it.

Ovando Hollister's 1867 comment, "At Idaho there is much good mining property, but the greater part of it is idle," could serve as an epitaph for the years that followed. Some silver found in the district brightened prospects as the 1860s ended, but it proved neither as rich nor as lasting as Georgetown's. Idaho Springs' mines were more gold-silver, instead of predominantly one mineral. This proved a decided advantage as the international price of silver continued to go downward, $1.33 in 1867 to 60¢ per ounce thirty years later. Keep in mind that the miner did not receive even this much, because the freighter and smelterman played the role of middleman between producer and consumer.

The lower Clear Creek mines, as they were called, made a slow recovery during McLean's early years in Idaho Springs; then, with the general collapse of silver mining, they came into their own in the 1890s. After the turn of the century, until the end of World War I (1918), gold production surpassed the declining county silver total. In the 1920s both gold and silver mining slumped, leaving behind the relics of another age, the vanquished hopes of another generation.

One of the great Idaho Springs mining projects of McLean's day was the Newhouse or Argo Tunnel, immediately east of town. The idea for it was similar to those planned at Georgetown, this one proving more successful. From 1893 into 1910, the tunnel was driven toward Central City, 21,968 feet. In 1913 a mill became part of the operation. The purpose was to provide economic transportation in bringing ore to the surface, drainage, and ventilation. It succeeded and operated long after McLean's death.

Having traced the vicissitudes of mining on both branches of Clear Creek, it becomes apparent that the industry was one of risk, physical labor, and a certain amount of luck. Mining is an industry unto itself, a fact most easily apprehended when one is underground in a world that seems unrelated to events occurring above it. Once into the darkness of a tunnel, shaft, or drift with only a lamp for a light (in McLean's day, a flickering candle), the intruder begins to feel cut off. The muted rumbles of distant ore cars, drilling, and blasting add to the detached feeling. Once the work area is reached, all is busy, noisy, and dusty; the rest of the mine echoes this activity. It is an intriguing world, one which can captivate the miner and convince him that there is no occupation to equal it. Even with the improvement of machines, mining in McLean's day continued to be dangerous, dirty, hard work. McLean tried to capture this world in photographs; others have tried verbally. It is a difficult task, either way, for anyone to portray it to the uninitiated.

In *Georgetown: Colorado's Silver Queen*, Liston Leyendecker quotes James Burrell, early-day Central City and Georgetown prospector and miner. Burrell, writing in the mideighties, recaptured the excitement and adventure of the early days. To understand this feeling is to grasp what kept mining moving at a feverish pace for decades.

> We went out and located twenty or thirty claims a day. We would find a crevice, put up our stake and then the boys would rush out and tack on to us at each end for perhaps twenty or fifty claims in length—not caring a shinplaster just so they joined end lines to the original discoverer. I am not sure that even the first locator had the vein—but we all staked on claims. We were all light hearted and hopeful. There were very few disputes and when one of the boys struck it, everybody was glad . . .

The prospector went out to search for the mineral; to him belonged the thrill of discovery, the anticipation of riches. After him, and occasionally the same person, as in Burrell's case, came the miner. Fossett pictures this in his *Colorado* (1879), quoting an unidentified issue of the *Colorado Miner:*

> For a period of about sixteen years, that enterprising individual, the "honest miner," has prospected and dug for the precious metals in our county with that energy and tenacity which is a distinguishing characteristic of "miner men," and, to some extent, is born of the circumstances in which he is placed. He has lived hard and worked harder, and with an undaunted brow has often faced the bitterest and sternest realities of life. He has gazed down the misty avenue of probabilities until what to others appeared to be the vague outlines of chance were to him all but an absolute certainty of his hopes. He has acquired fortunes with a few

weeks' or months' labor, and often, with a generosity bordering on recklessness, he has squandered it again in but little less time than it took to accumulate it. On the other hand, he has toiled incessantly for years without taking out a single "red," but his faith still continues unshaken and his perseverance unimpaired.. . . If he has not discovered the secret of the transmutation of metals, he has unfolded their rock-bound hiding-places, while mechanical and chemical science have sprung to his aid and rendered him indispensable assistance in their extraction. The brave and persistent miner has done all this and much more; not with the magic wand of an eastern fairy, but with a striking hammer, weighing from six to eight pounds, and other implements necessary to his vocation.

This unknown Georgetown writer knew of what he spoke, and the tendency to overdramatize in true Victorian phraseology can be excused.

The miner labored generally for eight to ten hours, starting not when he entered the mine but when he reached his work site, which might take as long as half an hour or more from the mine entrance. One shift would drill and blast, the next muck and tram, day in, day out. For this he was paid $2 to $4 per day, depending upon the year. Out of this pay the company would take $1 or so per day if the miner roomed and boarded at the company boarding house.

From the time he entered the mine, the miner placed his life in jeopardy. Few occupations involve more danger. The potential types of accidents are too numerous to list—blasting, rock falls, cave-ins, falls down shafts or ore chutes—to mention a few. Doctors' bills and hospitalization were responsibilities of the individual; no federal or company-sponsored programs helped the victim. One's fellow miners might donate some money, or the union might help out. Otherwise it remained for the injured party to pay his own bills and maintain himself during any recovery period. Insurance was nearly nonexistent; few companies would risk writing a policy on such a dangerous occupation. One reason miners joined a fraternal lodge was that they at least offered a burial policy! Because of the dangers and transitory nature of their occupation, many miners chose not to marry. Typically, however, the bachelors tended to drift on to the newer, booming districts, leaving those like Central and Georgetown, after the 1870s, with more married miners than normal.

Initially, most of the work was done by hand. Single and double jacking miners drilled the holes. In the former, a man holding a drill in one hand and a four-pound hammer in the other did the work; in the latter, two men worked as a team, one hammering, one holding. After blasting came the job of mucking broken rock into the ore car with a shovel. Finally the loaded cars were trammed to the surface, via the tunnel, or to the shaft and then out. All this involved physical labor. Adding to the effort was the work involved in timbering the mine and journeying in and out of a mine, generally located at a fairly high elevation. One begins to gain an understanding of the miner's workday. The only help he received at first was from mule or horse power to move the mine cars or to operate the whim that raised and lowered the bucket in the shaft.

Mechanical help came eventually. Georgetown and Silver Plume pioneered in the use of power drills in the late 1860s, using steam power. Better and more dependable power came with the introduction of compressed air, but miners were slow to adopt this machine, being a conservative lot. They also resisted those early drills because of the rock dust they generated, until a hollow drill was introduced which allowed water to cut the dust. McLean would be at Idaho Springs a decade or so before machine drills received general acceptance.

Major advances came in the form of power-operated cages that replaced the horse whim, trams that carried ore from mine to mill or railroad, and the introduction of electricity to the mine and mill. Not only did electricity bring a new form of power, it also provided a much better source of light and improved working conditions. The impact on the mining communities was equally as great. Better and more powerful pumps helped keep the water level down, and engines generally replaced the horse or mule to haul ore carts and transport men faster into and around the mine. All this McLean would see.

Still the miner labored and worked under perilous conditions. Safety legislation came very slowly. So strong was the opposition to the eight-hour concept that a decade of legal maneuvering and management-vs.-labor conflicts resulted when it was proposed legislatively. So steady the hold of management during McLean's lifetime that it successfully beat back the threat of any curbs on its "rights."

Management was aided by a major industrial trend, corporation control, which had been noticeable since the days of placer mining at Gregory and Jackson Diggings. The need for money, the sale of stocks, the increasing complexity and cost of mining, all militated against the small operator, even in Clear Creek County. As was shown in Georgetown, consolidation was the result, and absentee ownership became the normal condition. An industry that had once gloried in its individualism became for most miners little different from a day wage job in any steel mill or railroad yard. Their independence was gone, their opportunity to make a stake and discover their own bonanza claim as lost as the frontier era which fathered the image. No new districts opened after the early nineties to which they could migrate; the miners were trapped. The miners of the 1900s retained their pride in their occupation and their skill. Otherwise they had little in common with their predecessors of the 1860s.

To be sure, they opposed the impersonal corporation control by joining unions. Colorado's political and economic climate was not conducive to unionism, however, and what followed was a decade-long fight that began in 1894. It eventually ended with the Western Federation of Miners being broken as a spokesman for Colorado hardrock miners. Georgetown, Idaho Springs, and Central City played only a small role in this fight, evidence of their reduced significance. A strike that broke out at Idaho Springs in May 1903, over the eight-hour day, resulted in the dynamiting of a mine building and the subsequent deportation of union miners. The real battle, however, was waged at Cripple Creek and Telluride, both bonanza mining districts in the 1890s and after.

McLean had observed nearly the whole nineteenth century saga of Colorado mining, missing only the early years. Evidence even of them could still be seen, when he arrived, in remnants scattered about the valley, and experienced by listening to the stories of those who had been there. Certainly the Leadville and Cripple Creek rushes re-created the chaos and adventure of the 1859 Pike's Peak rush, though slightly updated by changing times. McLean was fortunate to have lived through these years, and we are fortunate that his photographic record of it has been at least partially preserved.

To return once more to James Grafton Rogers, who looked over the change that had come to his mountain valley and wrote in the late 1960s,

> The mine tunnels are closed today with iron gates or blocked with fallen rock. There are no miners seen with dirty boots and tin lunch buckets. The streams flow clear. Even the language of the old mining days—full of words like stope, double jacks, gads, moils, and winzes, which were once known in every household—is no longer used or understood.
>
> Something of the great mining days remains. There are scars on the mountains. There are elderly men and women here and there who sit under electric bulbs at the same tables where kerosene lamps once burned and who will talk of fortunes made, or workers who died from "miners' consumption" and about unanswered prayers for a lucky strike.

Statistics
Mining Production

Clear Creek County

Silver		
1866	$ 20,000	
1870	481,000	
1875	1,666,000	
1880	2,188,000	
1885	1,450,000	
1890	1,910,000	
1895	1,030,000	
1900	842,000	
1905	422,000	
1910	256,000	
1915	199,000	
Gold		
1859-65	2,000,000	(estimated)
1902	930,000	(peak year)
Lead		
1890	541,000	(peak year)
Copper		
1917	155,000	(peak year)
Zinc		
1916	344,00C	(peak year)

Total production 1859-1923: $87,191,220

Gilpin County

Gold		
1859	$ 241,000	(estimated)
1862	1,161,000	(estimated)
1865	1,451,000	
1870	2,120,000	
1875	1,395,000	
1880	2,380,000	
1885	2,051,000	
1890	805,000	
1895	1,196,000	
1900	1,655,000	
1905	1,450,000	
1910	687,000	
1915	562,000	
Silver		
1885	321,000	(peak year)
Lead		
1892	89,000	
Copper		
1902	193,000	(peak year)

Total Production 1859-1923: $98,411,325

Throughout these years the price of gold was approximately $20 per ounce; silver, lead, copper, and zinc varied from year to year. Production figures at their best are often educated estimates.

For comparison, Teller and Lake counties' total production figures were, respectively, $324,000,000 and $425,000,000.

Source: Charles Henderson, *Mining in Colorado*

Photographs

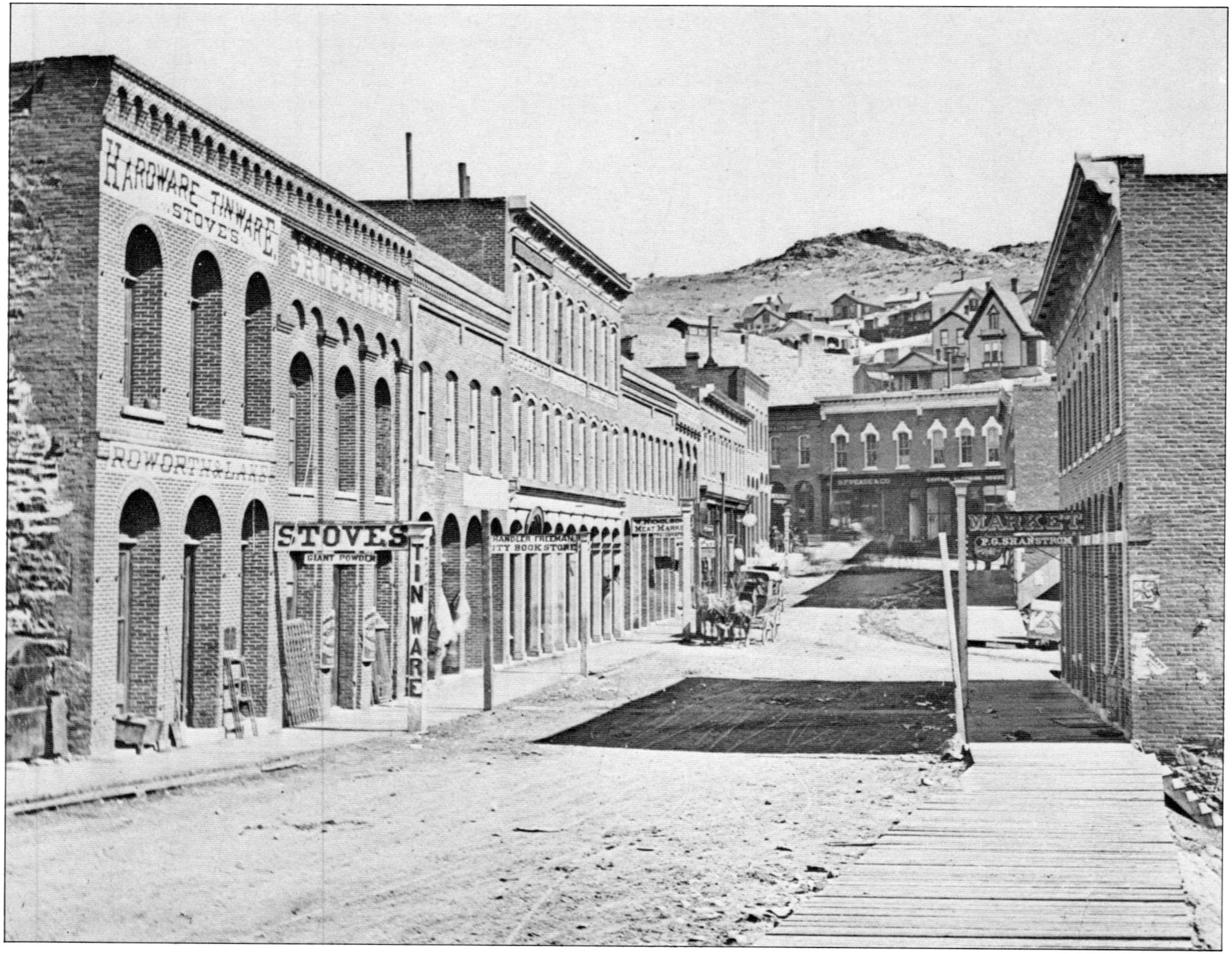

1. A quiet morning on Main Street—Central City sometime after the monumental fire on May 21, 1874, and before 1877. Most of the center of town was destroyed but was quickly rebuilt.

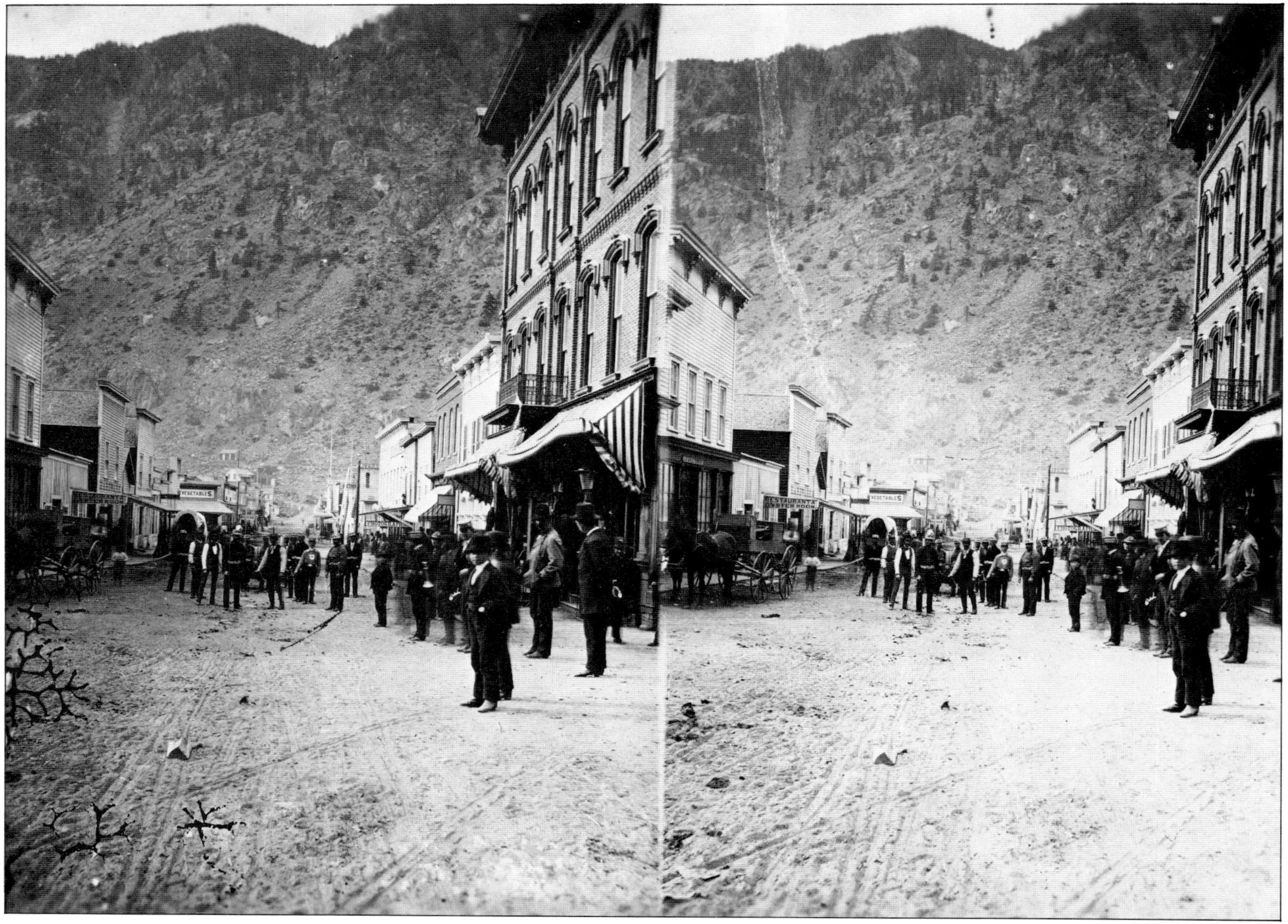

2. Stereoscopic view of Georgetown about 1877. This was Alpine Street, now Sixth Street. On the right of the street can be seen a sign, "Photographs," one of the early studios that McLean opened in Colorado. One of Georgetown's famous fire companies poses in the middle of the street; in the 1870s, there was none finer in Colorado. A practical necessity, the companies also were social organizations and a source of town pride when they defeated the boys from Central, Denver, and Boulder in tournaments.

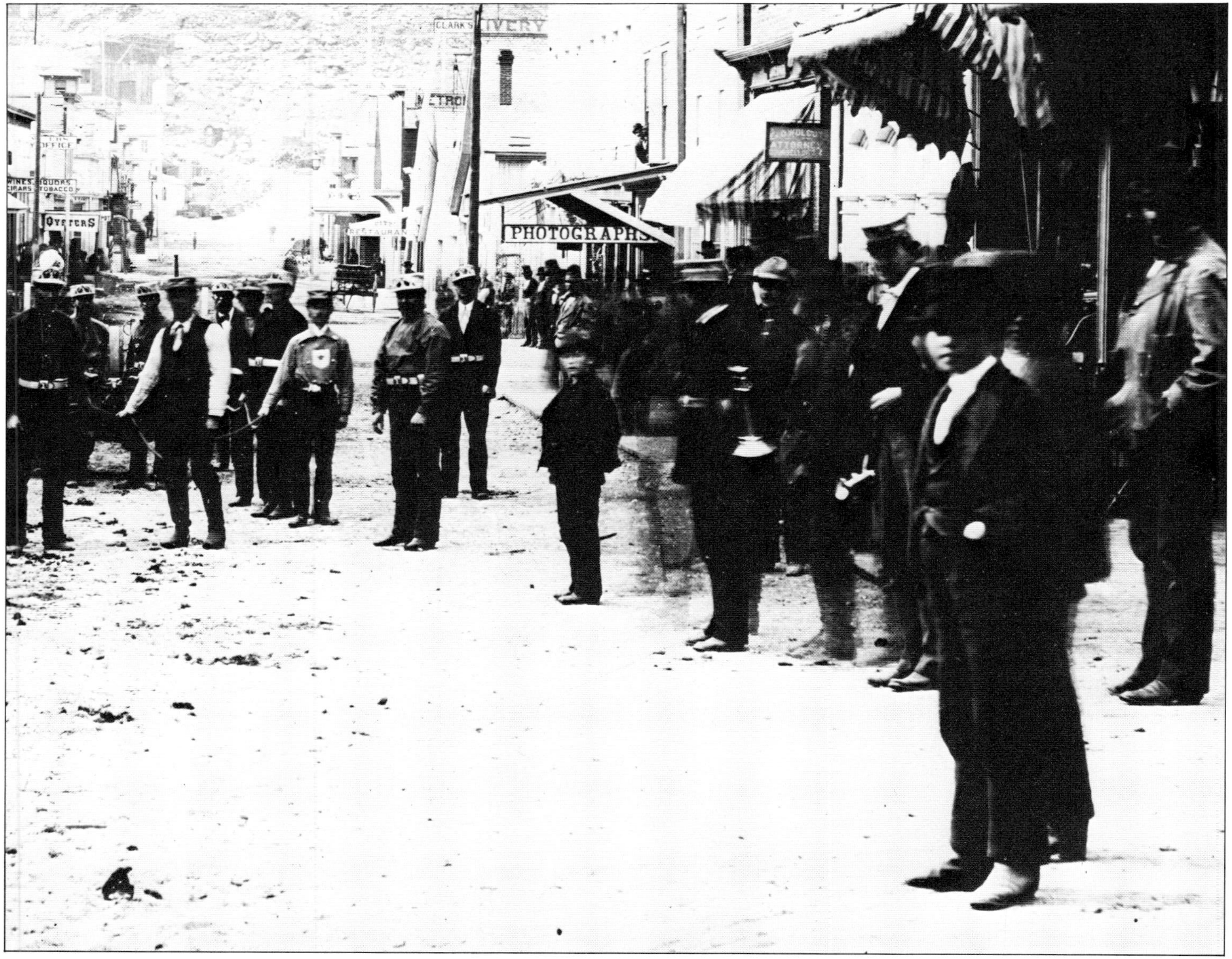

3. McLean's studio was surrounded by the typical businesses of a mining town of Georgetown's prosperity and size. Edward Wolcott, attorney and counselor, whose sign hangs here, eventually became a United States Senator. Brick buildings were favored for the substantial appearance and air of permanence they gave, and as fire breaks.

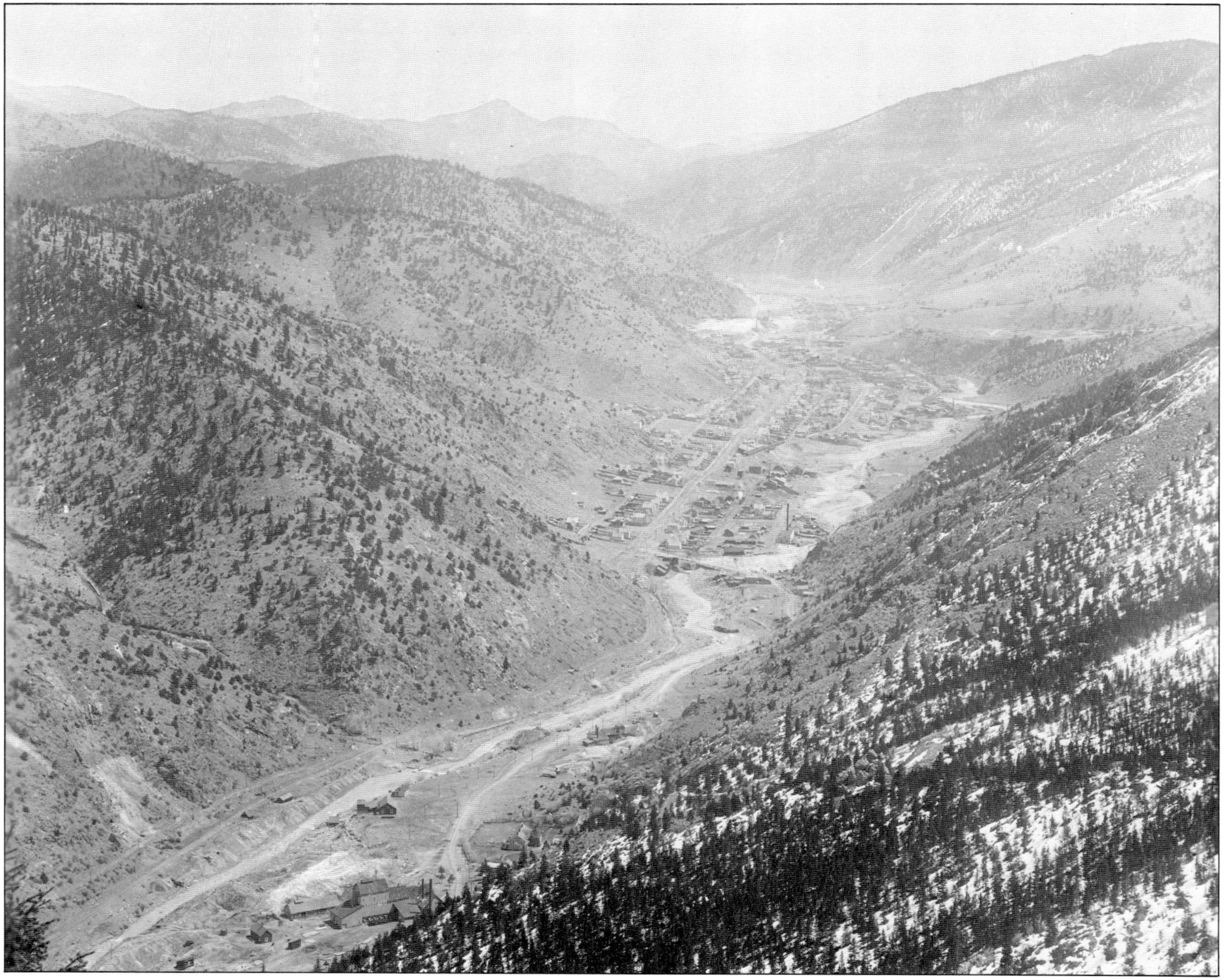

4. Nestled in the valley is Idaho Springs. Here, in January 1859, gold was discovered. This picture was taken from Bald Mountain, looking east. The Stanley Mine is in the foreground, and the Argo Tunnel and Mill are found in the light oblong area in the top center.

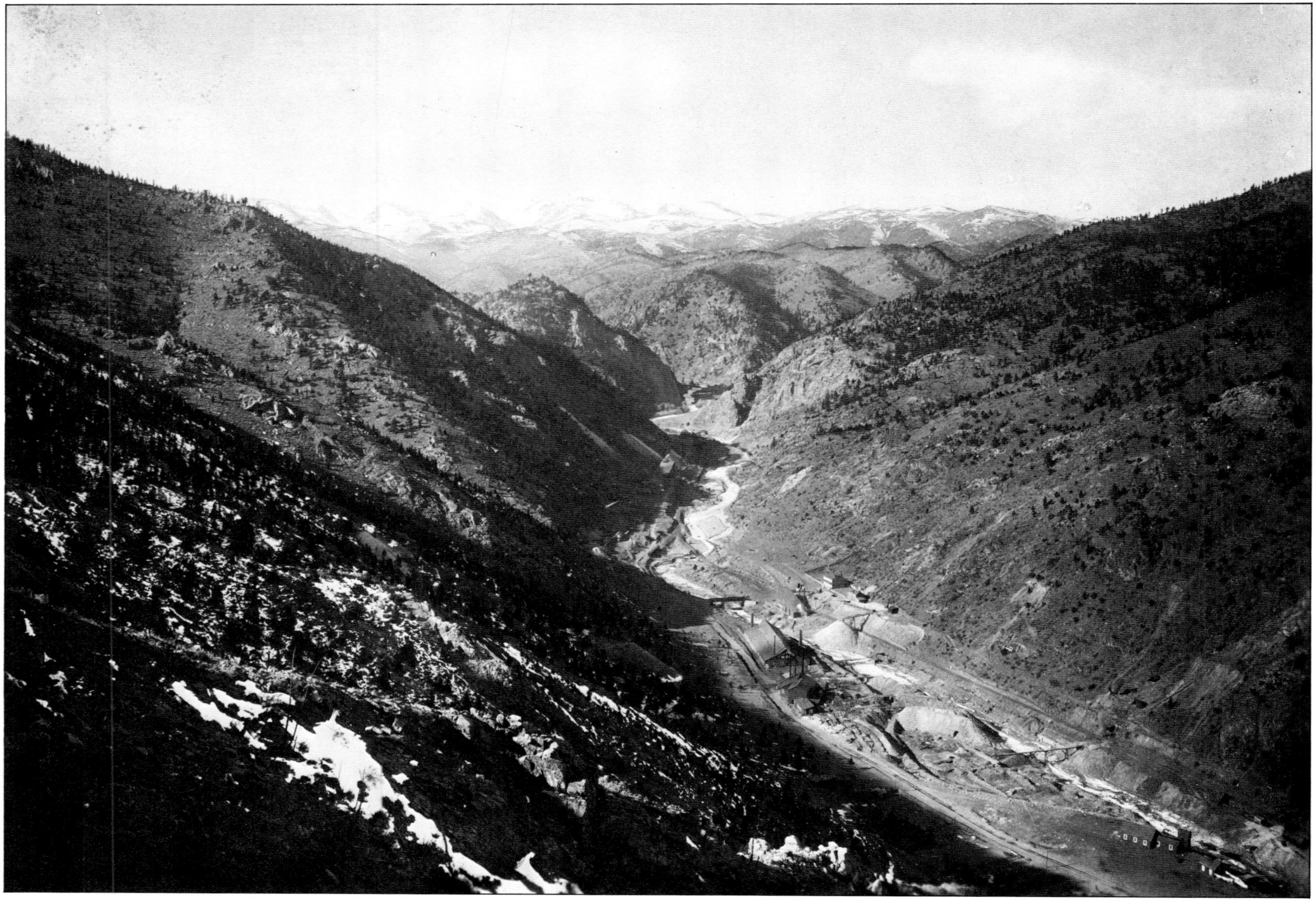

5. Clear Creek Valley looking west from Idaho Springs, which sits out of sight to the right. Except for the absence of trees, it appears much as the 1859'ers must have discovered it. The prying and digging of a generation of miners is evident throughout.

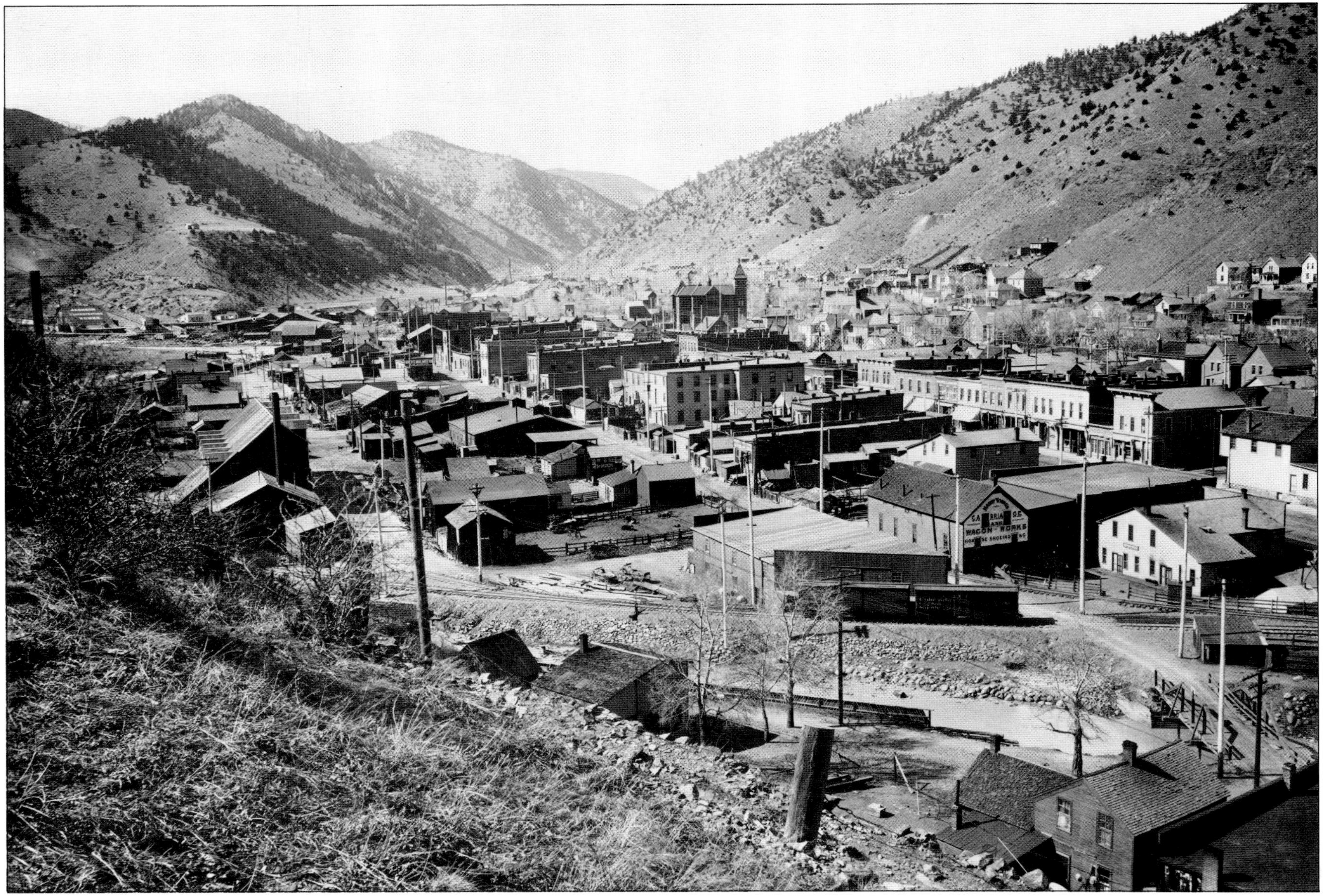

6. Idaho Springs at its turn of the century best. This is the main part of town, showing the business district. Opportunities for extracurricular pleasure were found in the row of houses at the bottom. Also a business district of sorts, this was Water Street, the red-light district.

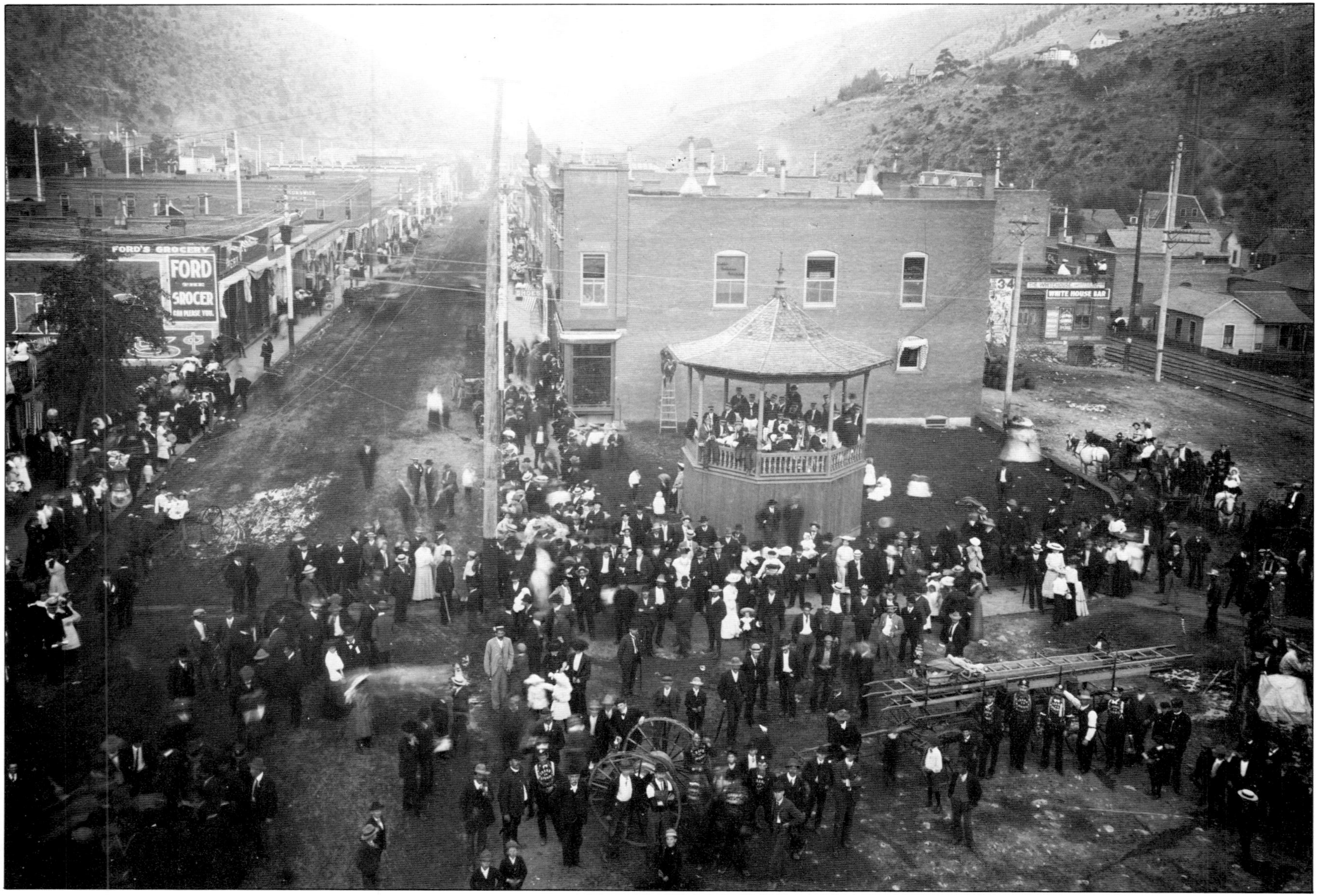

7. A warm summer day was a time for celebration. The Idaho Springs Fire Department is probably sponsoring a show or contest, and the band plays in the pavilion for the music lovers. The flags suggest this may have been July 4.

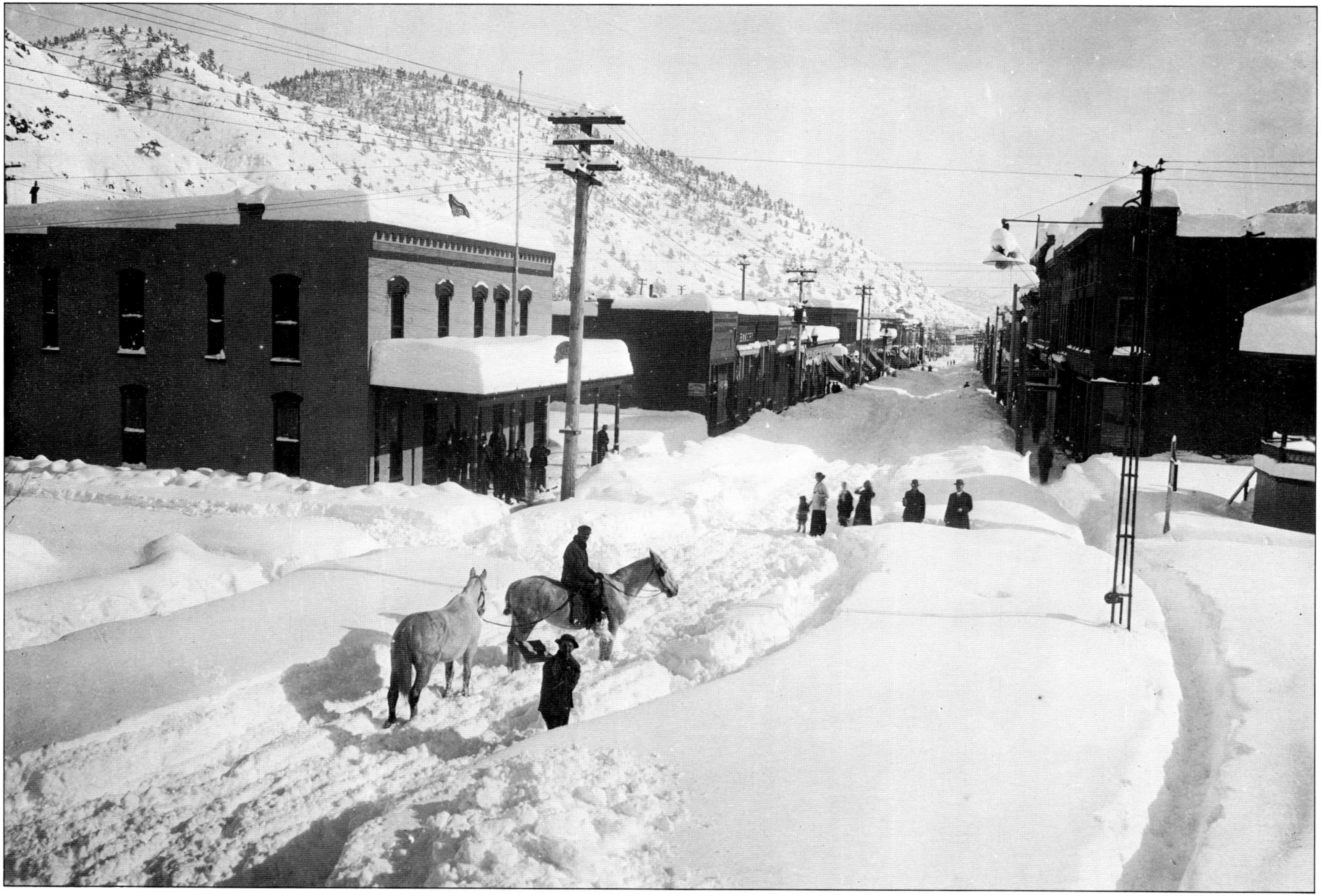

8. The same location as the previous photograph, this is the corner of Fourteenth Avenue and Miner Street in a different season of the year. Folks are digging out after the "big blizzard" of 1913. Electric lights now brighten the streets, and some of the businesses have changed.

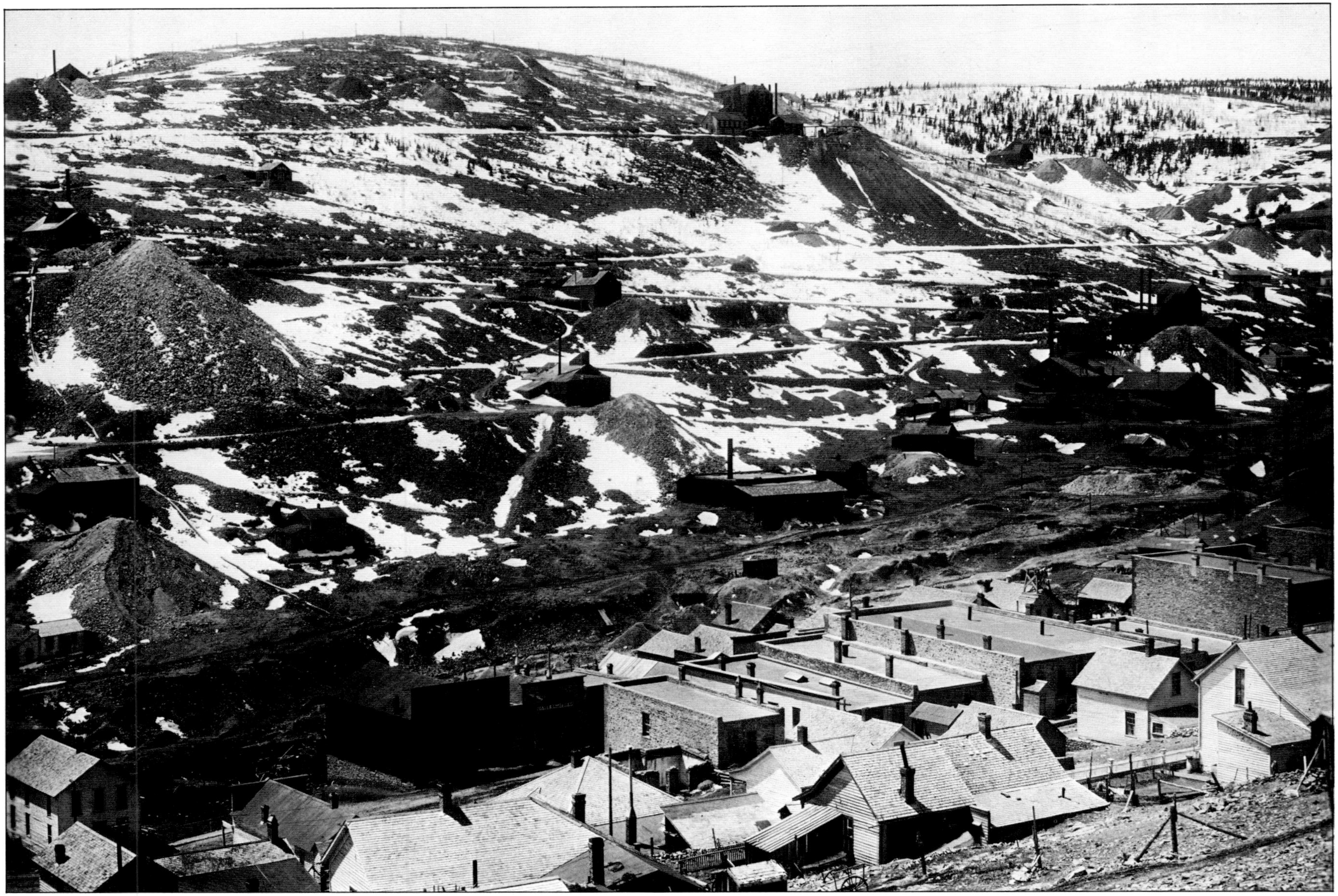

9. On trips from his studio, McLean visited nearby mining camps, few more interesting than Nevadaville, a community above Central City. Nearly gone today, it once hummed with activity. In the winter, a snow run between here and Black Hawk was famous among all sled owners.

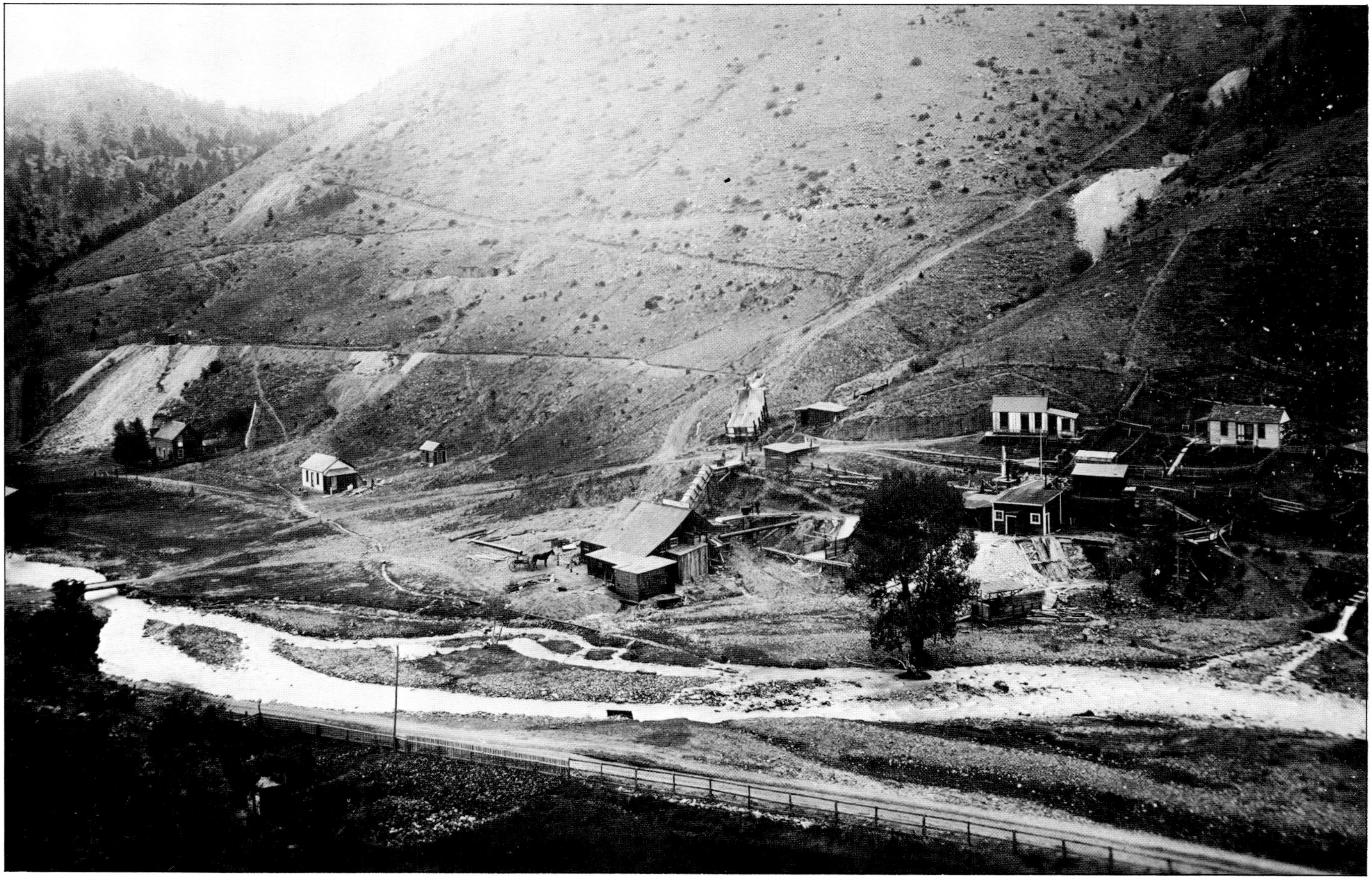

10. Some mining settlements were simply scattered buildings around a mine or mill. The little white building with several children in front was the Chicago Creek school house. Right across the street sat the Egan Reduction Company. A sharp eye can detect two miners and ore cars above the mill, and a third man standing above the empty loading bin. What the windmill was used for is unknown.

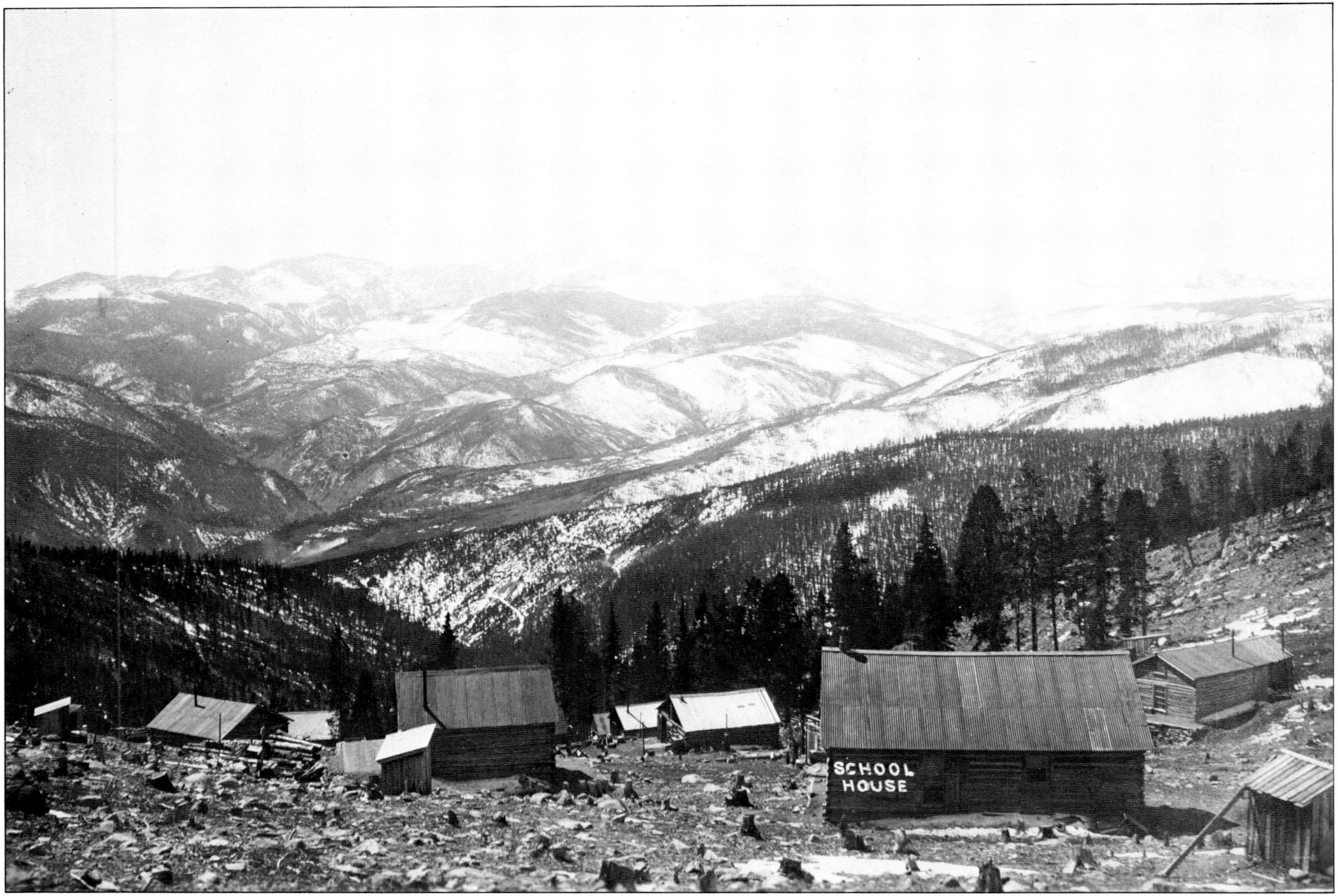

11. "How Fleet the Works of Men"—Lamartine has long since been abandoned. The little community hit its peak in the late 1890s, and the houses were built like steps down the mountainside. No one could fault the view, but winter in these log cabins must have tested one's endurance.

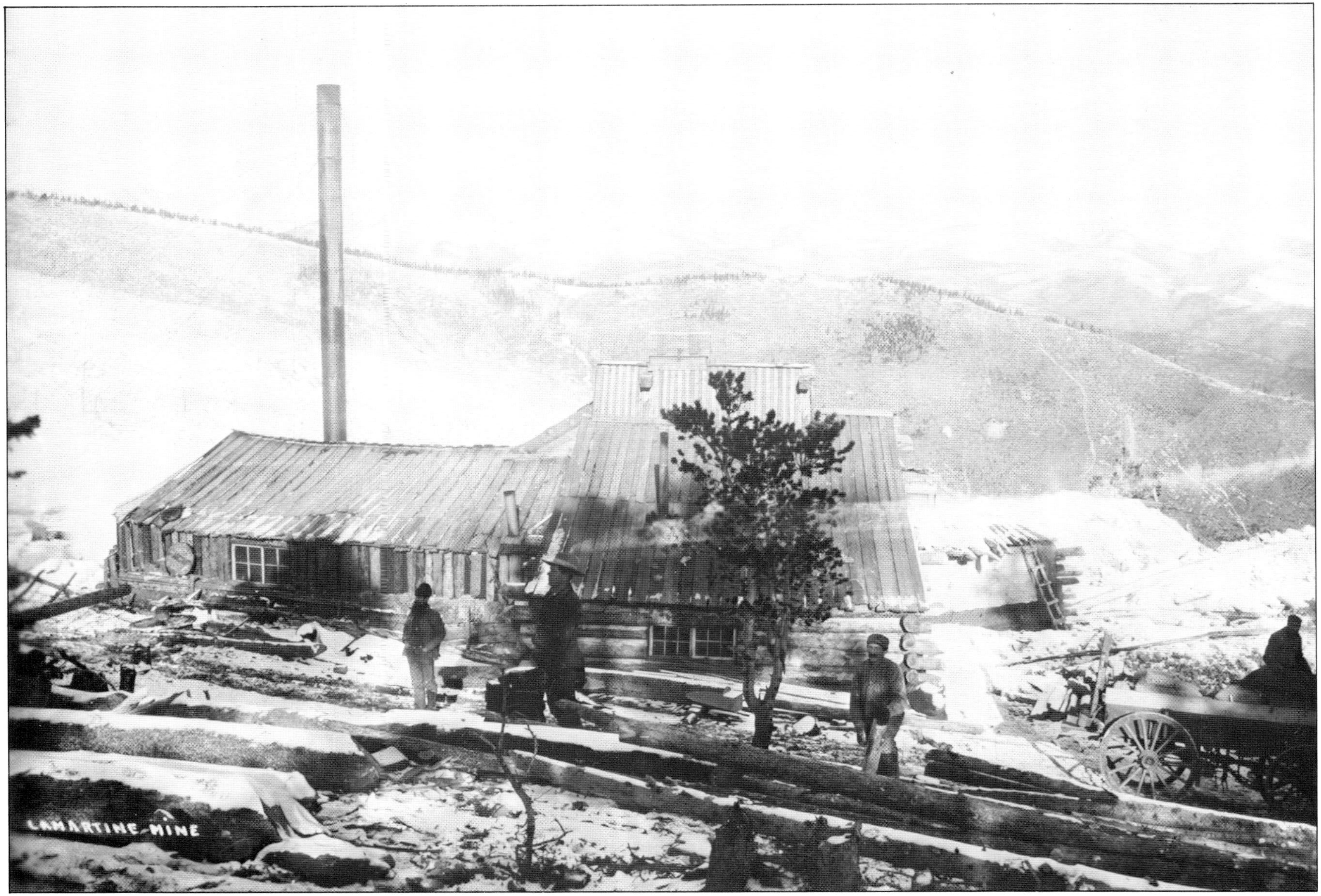

12. High, windy, and seasonally inaccessible, the Lamartine Mine provided the foundation for the camp of the same name. Wagons freighted supplies in and ore out; simply bringing in fuel must have been a chore. The debris seen around this shafthouse was typical.

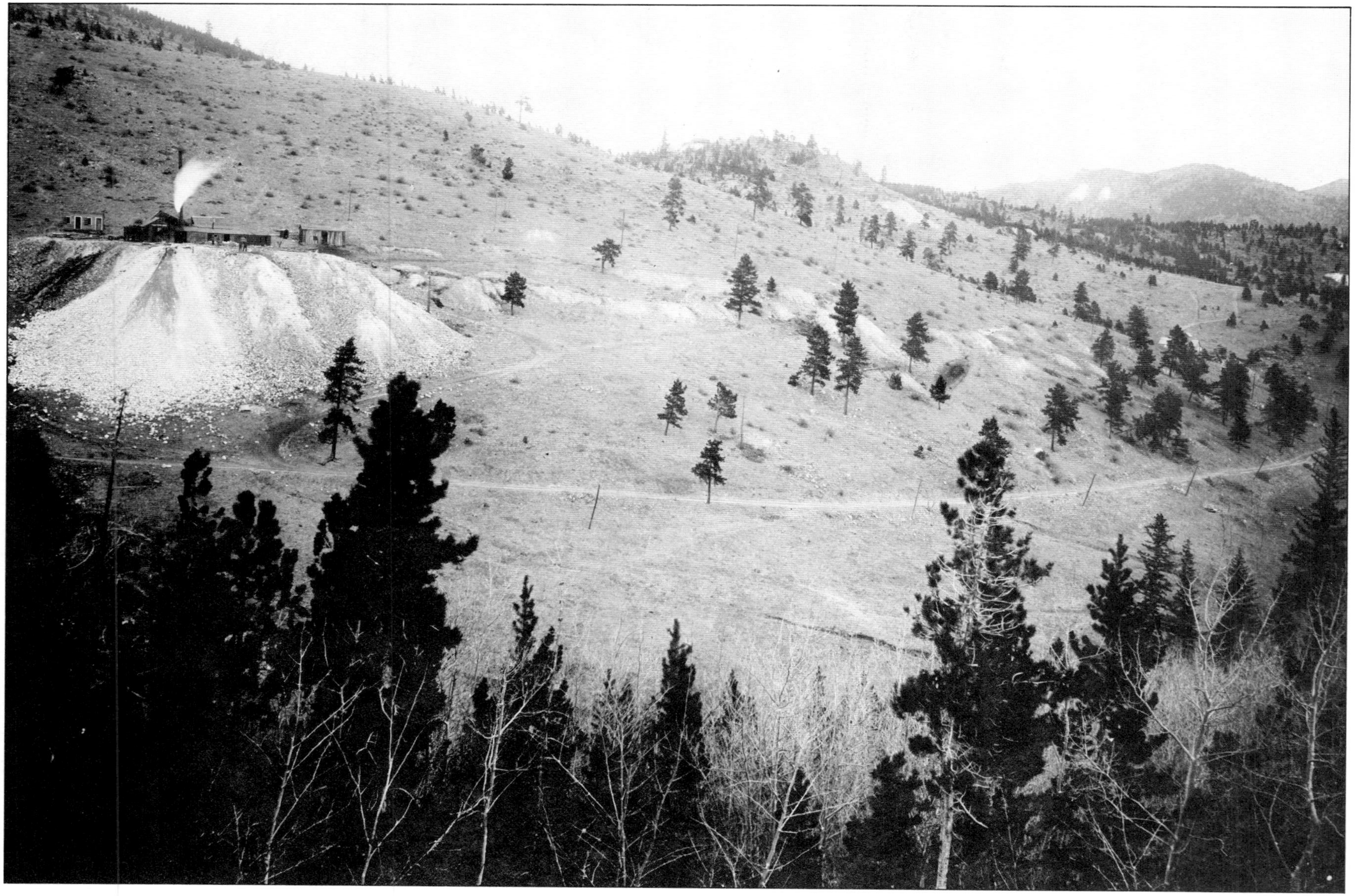

13. McLean spent much of his out-of-the-studio time photographing mines and mine buildings. The Fraction Mine was operating on this day. The rest of the "blast and prayer" holes on the hillside indicate less success. The power lines stretching hither and yon show the acceptance of this form of power on a 1908 or later date.

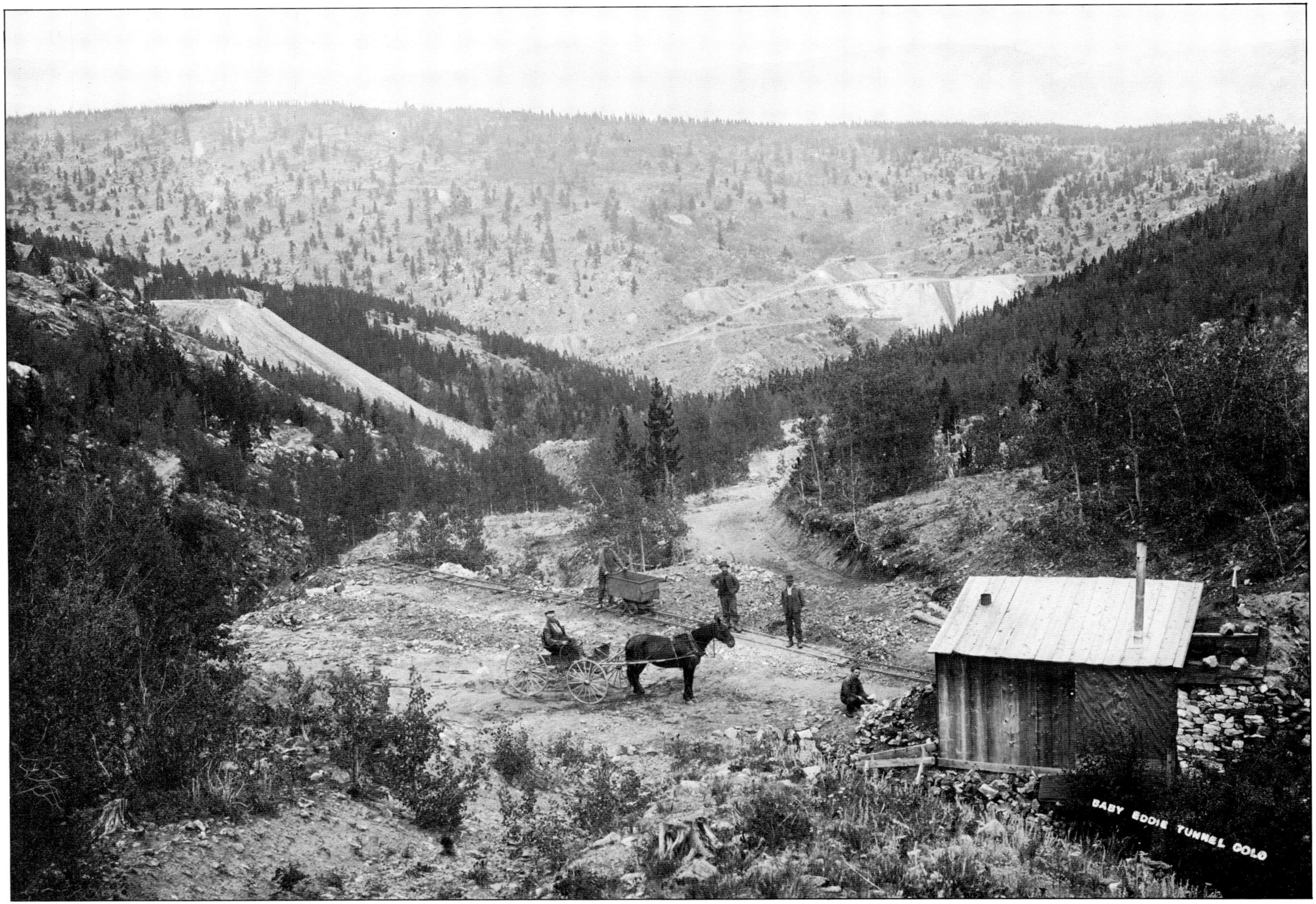

14. A photograph was more likely to be sold if people were in it, hence the entire crew, and maybe the owner as well, appear here. The Baby Eddie Tunnel was in its infancy. The mine dumps to the left and across the valley indicate large-scale operations.

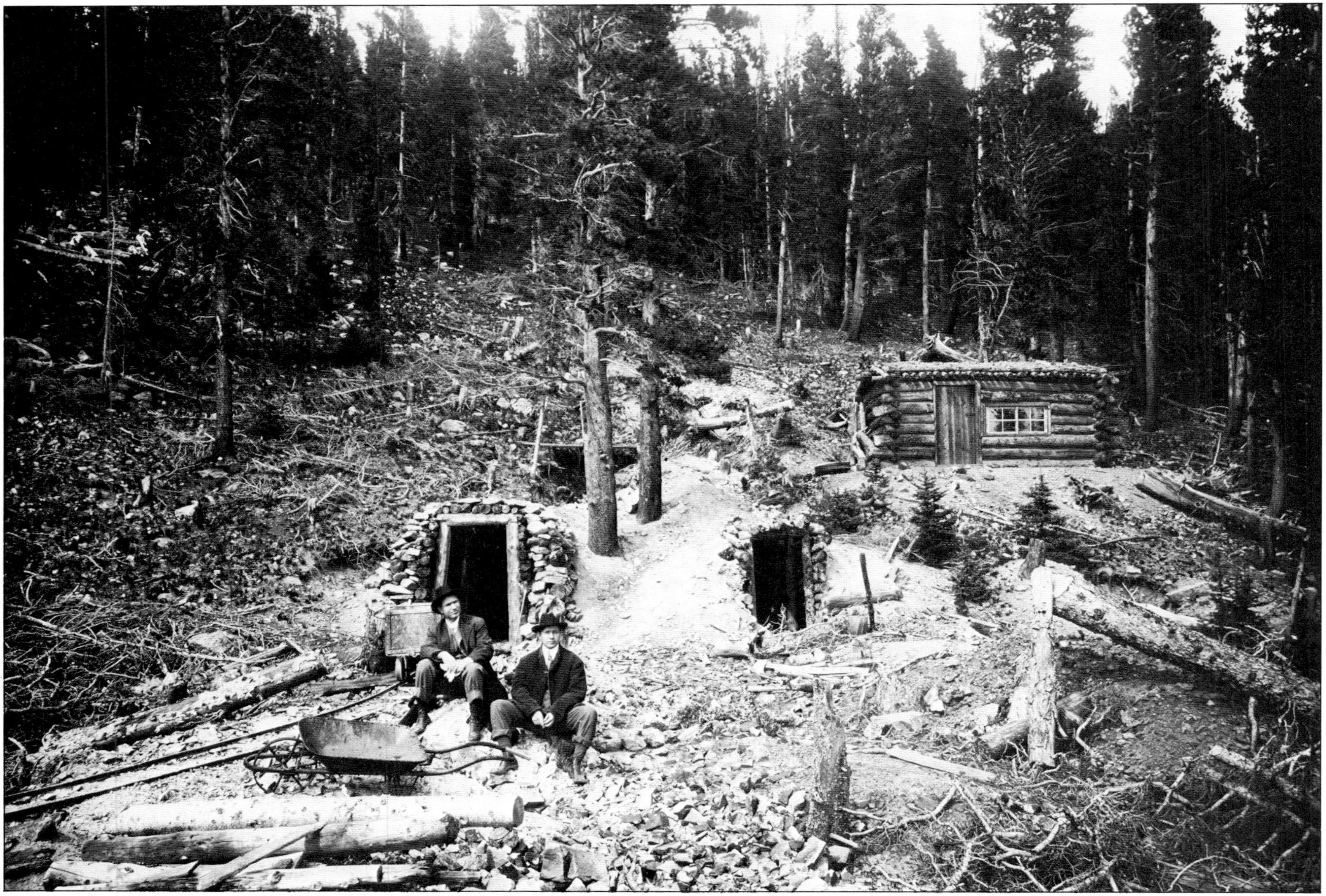

15. Most Clear Creek mines were small operations, maybe not as small as this, however. The owners, if that is who these men are, dressed up for McLean. They apparently had just started to work their prospect and were tunneling along the vein.

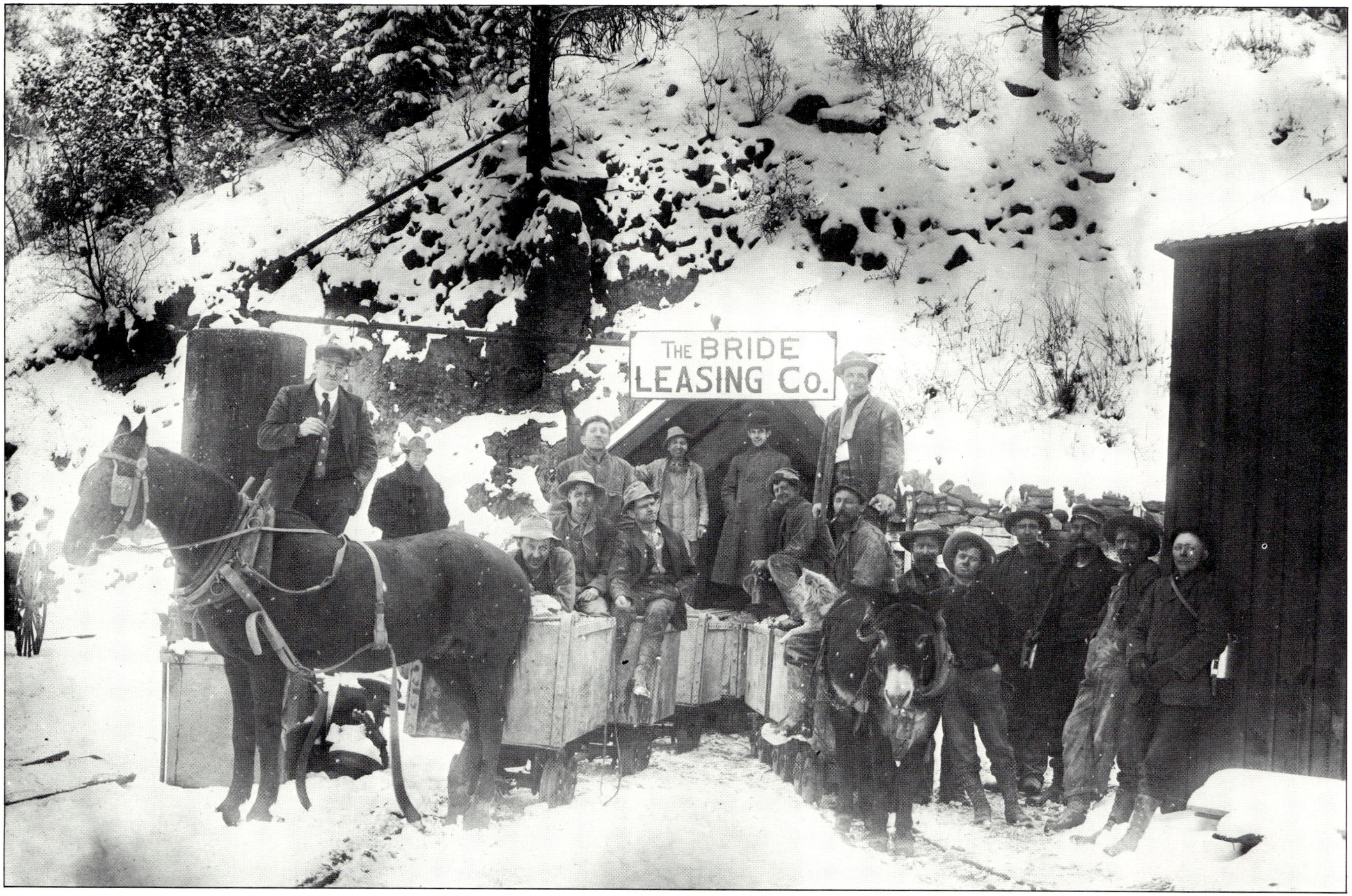

16. The Bride Leasing Company turned everyone out, from dog to mule to man, for this picture. The crew appears ready to go to work. Leasing was common when the owner did not have the experience, resources, ambition, or desire to work the property. It provided a way for an ambitious miner to become established and become an owner eventually.

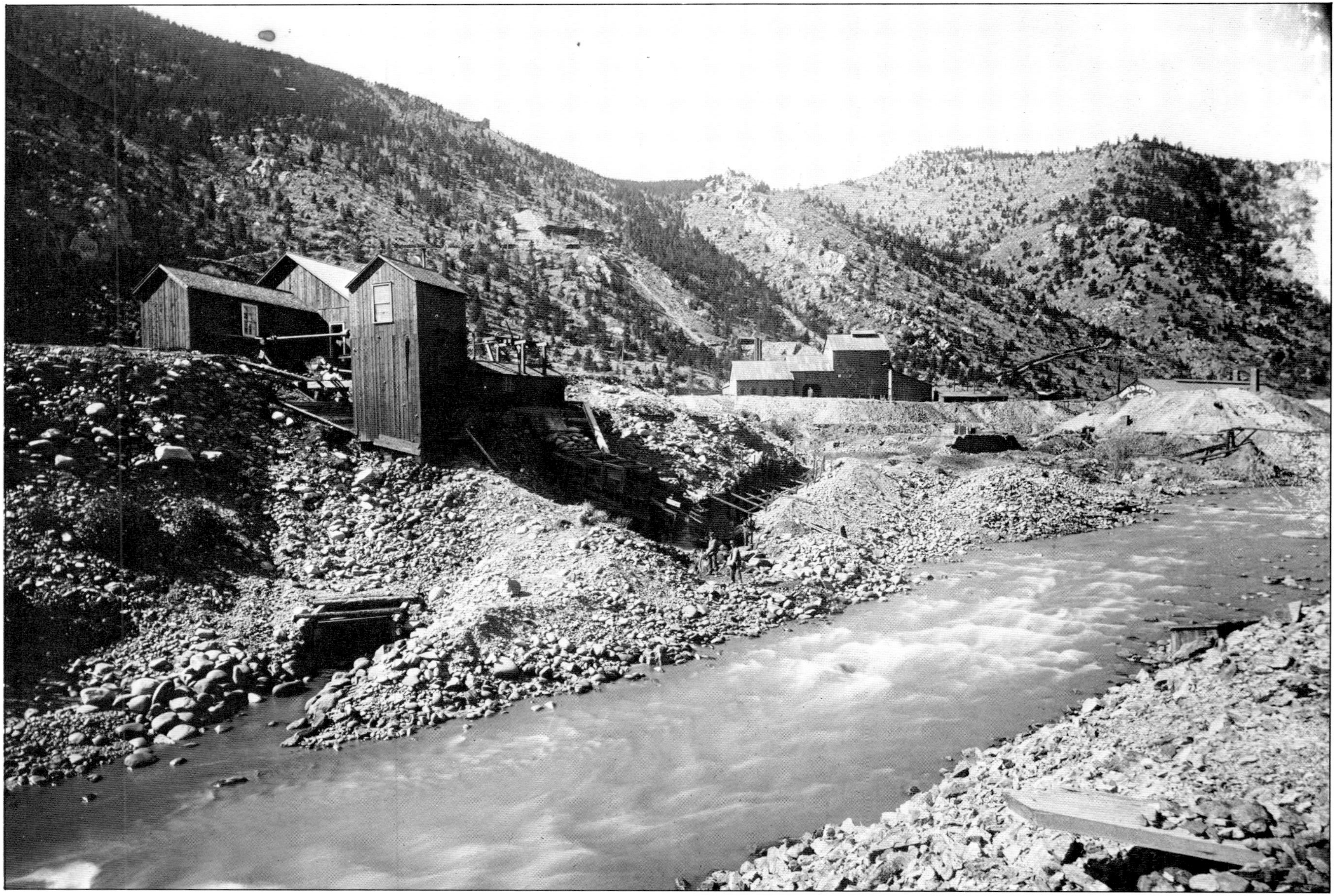

17. Idaho Springs provided its share of larger mining operations, including the Stanley Consolidated Mines, just west of town. As late as 1978, it was still being operated on a part-time basis, an extremely long life for a mine.

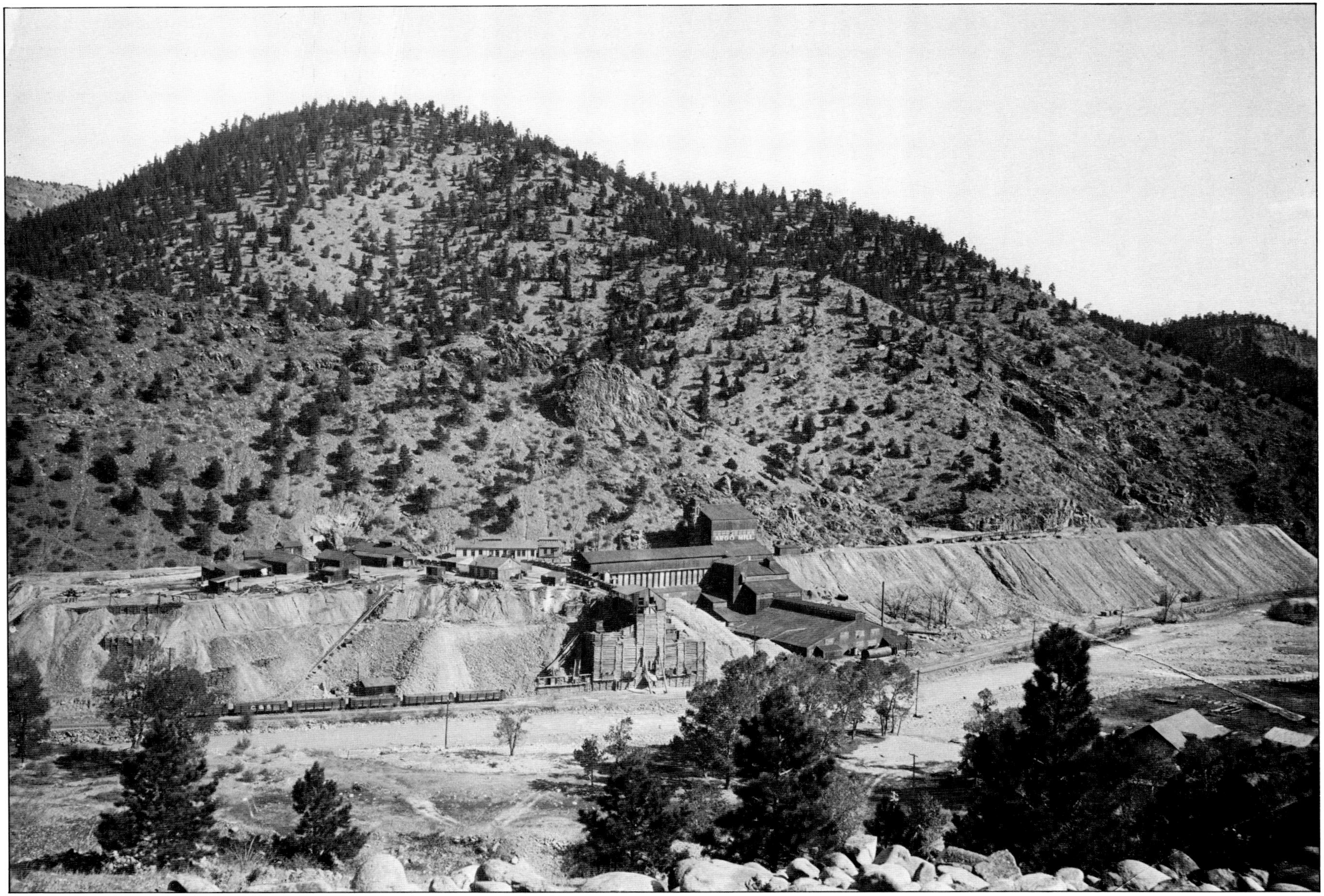

18. The Argo Mill and Tunnel after 1912. Active until 1943, when an underground flood closed the tunnel, the Argo operation helped keep mining going nearly all the way to Central. Famed western mining man Samuel Newhouse started the project in 1893.

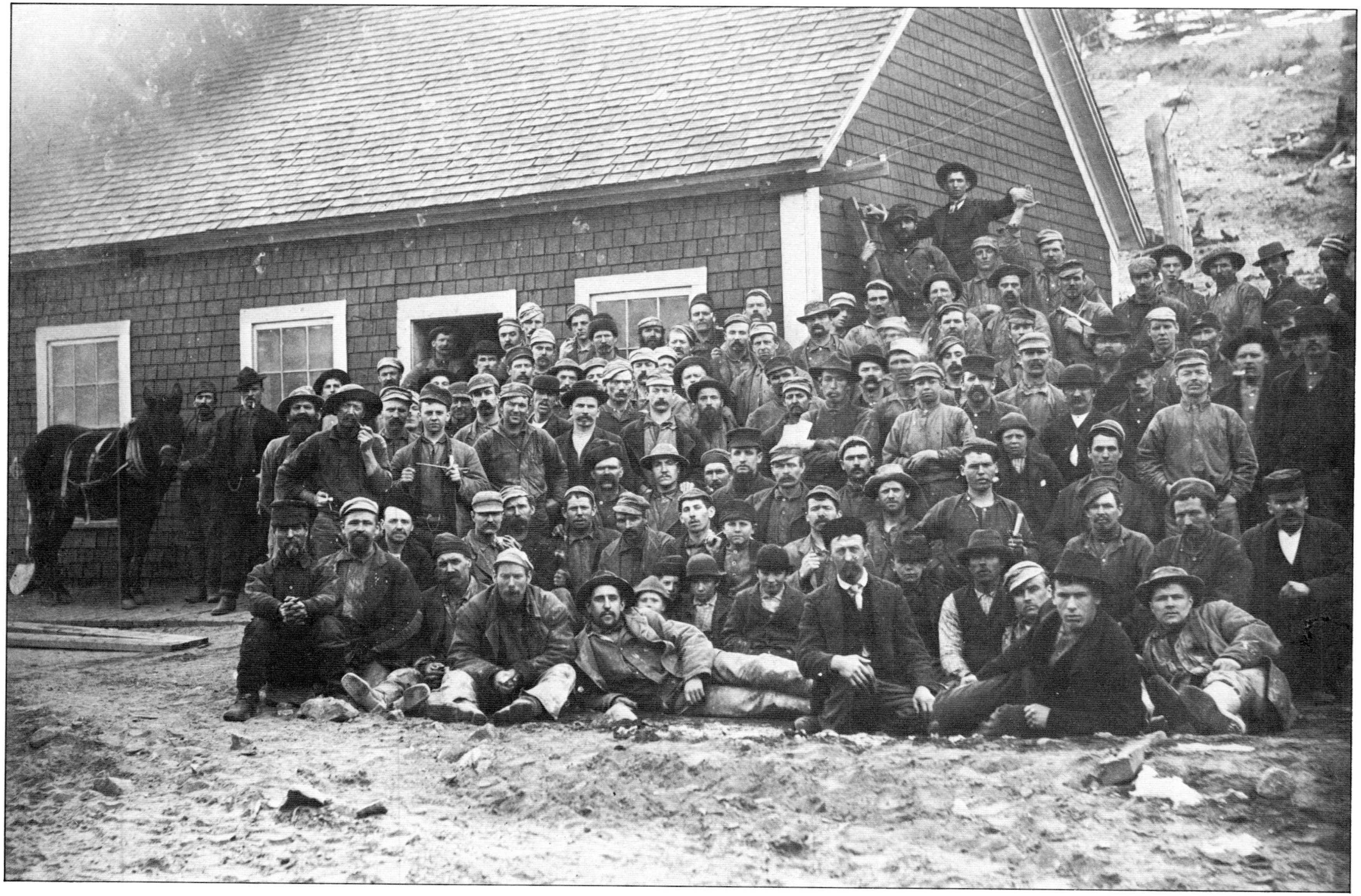

19. The entire above- and underground crew of some unidentified mine, including one faithful mule in harness. The young boys either worked on the surface or came to have their pictures taken with their fathers. Note the absence of hard hats and the varied styles of battered felt hats, probably stiffened with resin. The lack of oilskins indicates this to be a dry mine.

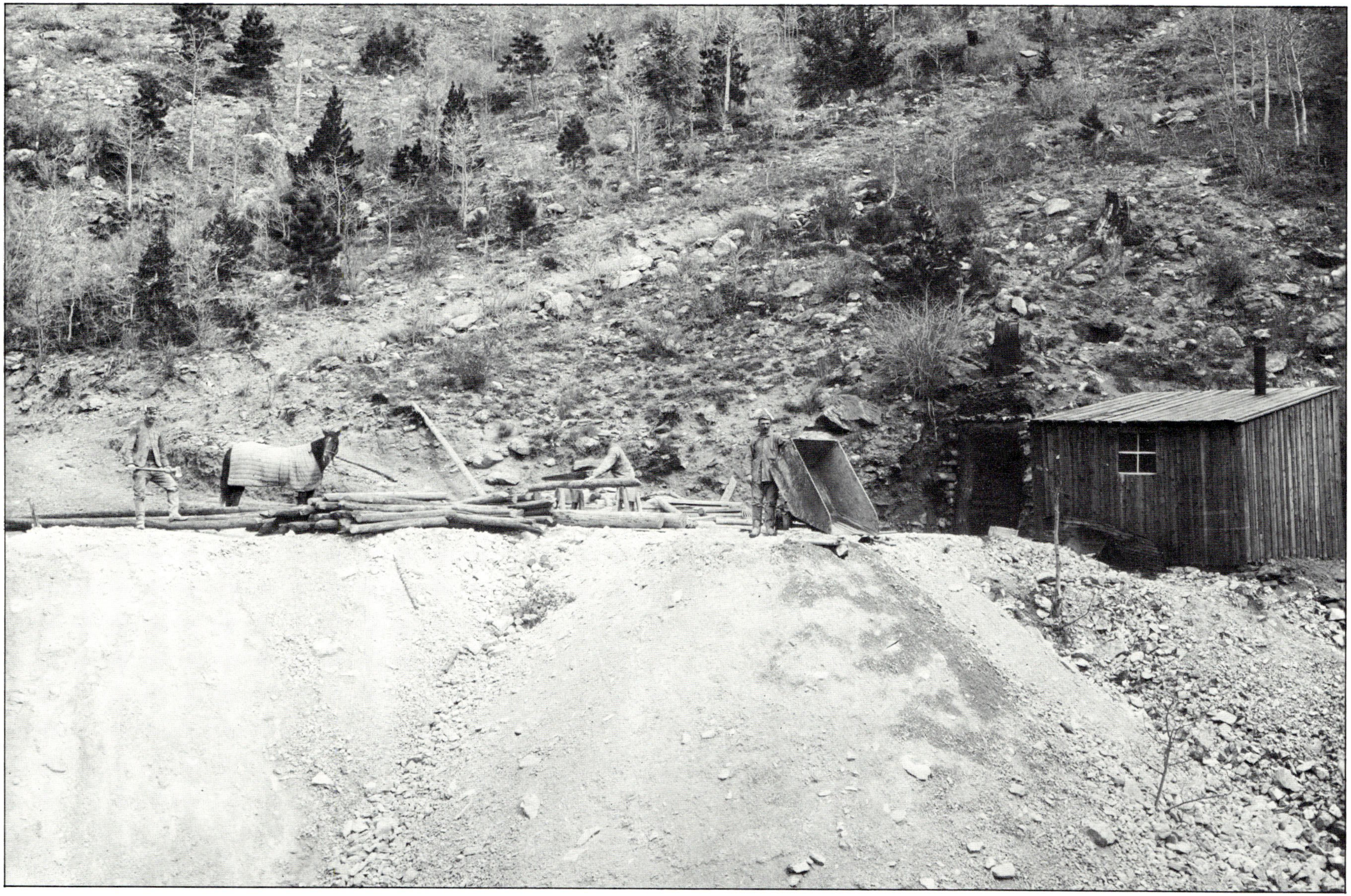

20. Part of the essential work of a mining operation was cutting timbers to shore underground workings and timber shafts. To strengthen weakened areas and prevent cave-ins took acres of timber in some large operations, where several timbermen worked to supply the miners' needs. In this small mine, miners themselves are cutting and cleaning the logs.

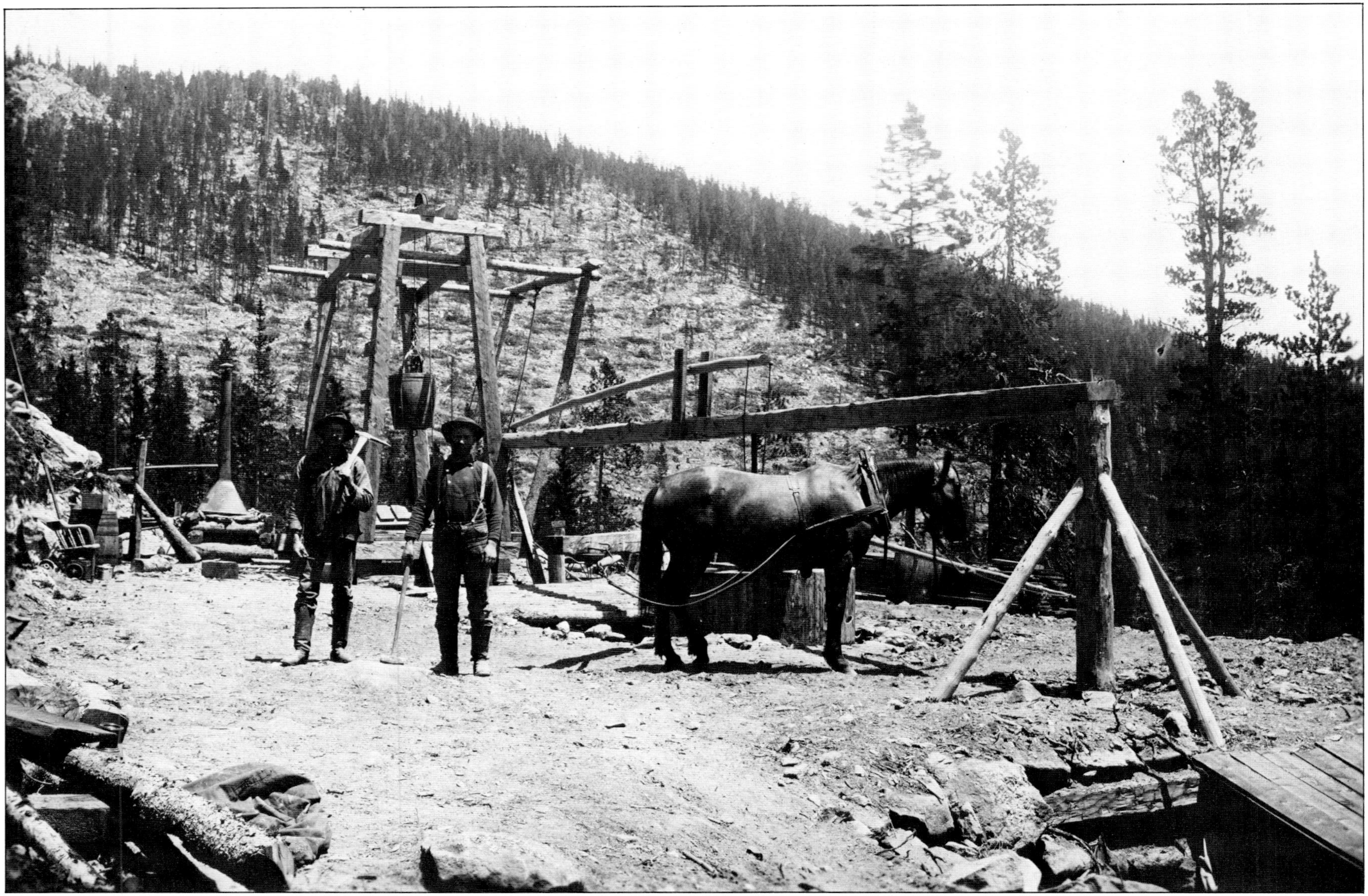

21. A "one horsepower" whim is used to raise the bucket in the background. The horse, working in a circle, rotated a drum and raised a bucket to bring up miners or ore. A whim was used early in Colorado, replacing the hand-operated windlass. It, too, was eventually replaced by the steam hoist in all but the very smallest operations.

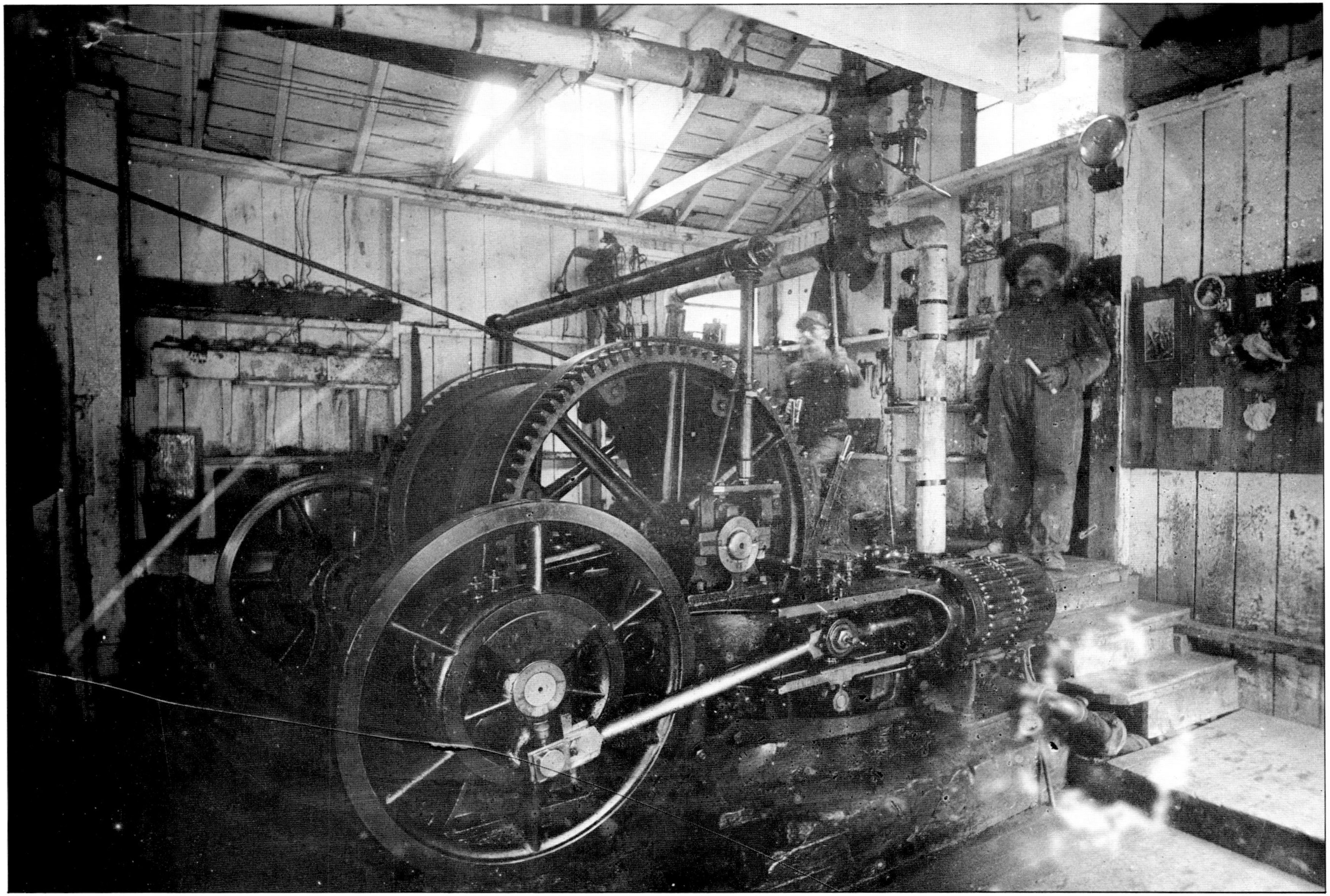

22. The steam hoist raised and lowered the bucket or cage in the shaft. The hoistman, hoist engineer, or cageman—by whatever name he was called—held one of the most critical jobs in the mine. The bell at the right rang to signal what was needed and where. Note the "pinups" at the right.

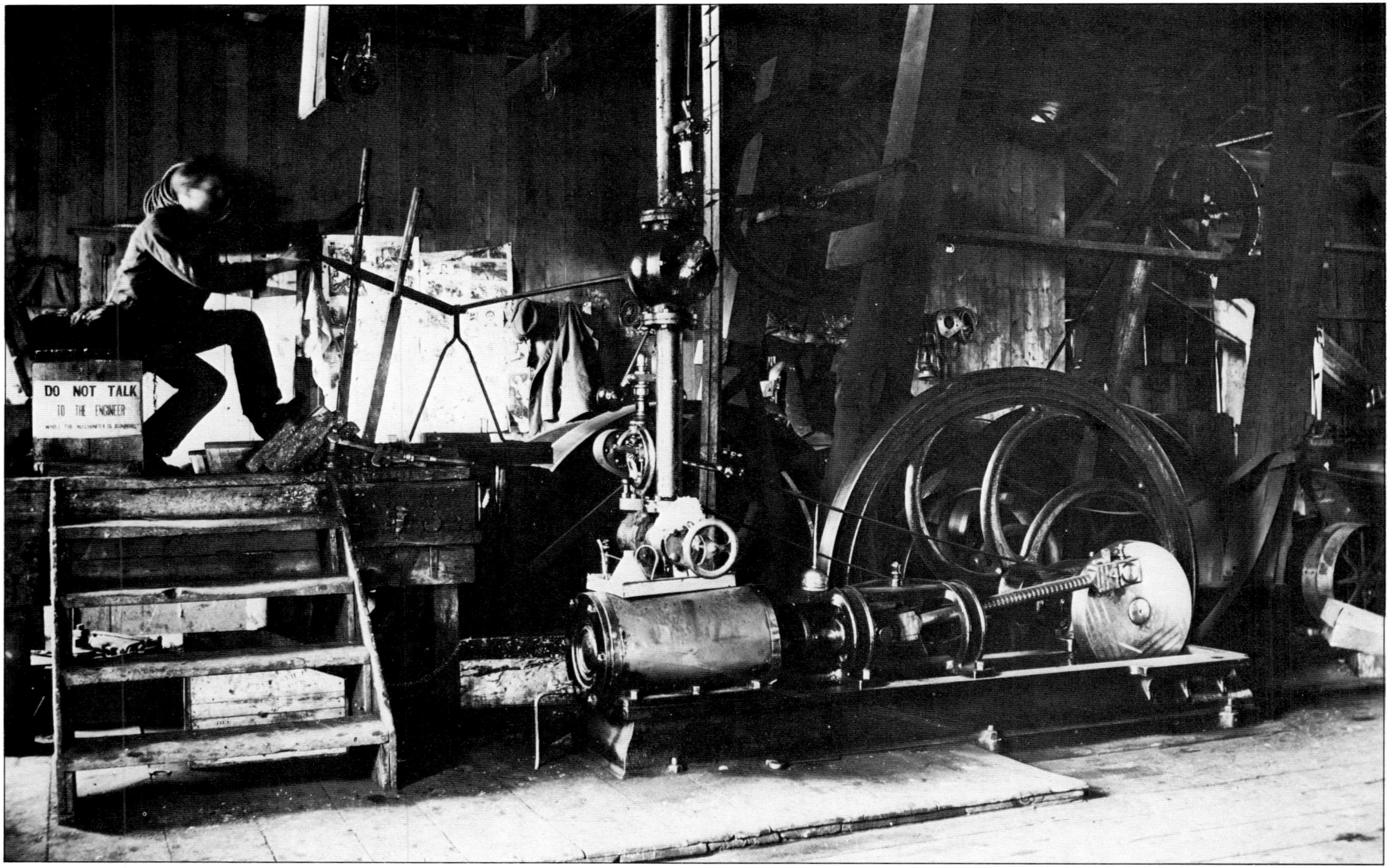

23. From somewhere down in the mine the signal came and the engineer responded. Operating the hoist was a life-and-death matter, since anyone in the cage rode on the engineer's skill. It was essential that no one talked to him while the machinery ran, and that he did not drink on the job or report to work in any condition other than cold sober. These steam hoists were beautiful machines, all brass and steel and carefully machined. The pictures from the *Police Gazette* in the background kept the engineer entertained during slow times.

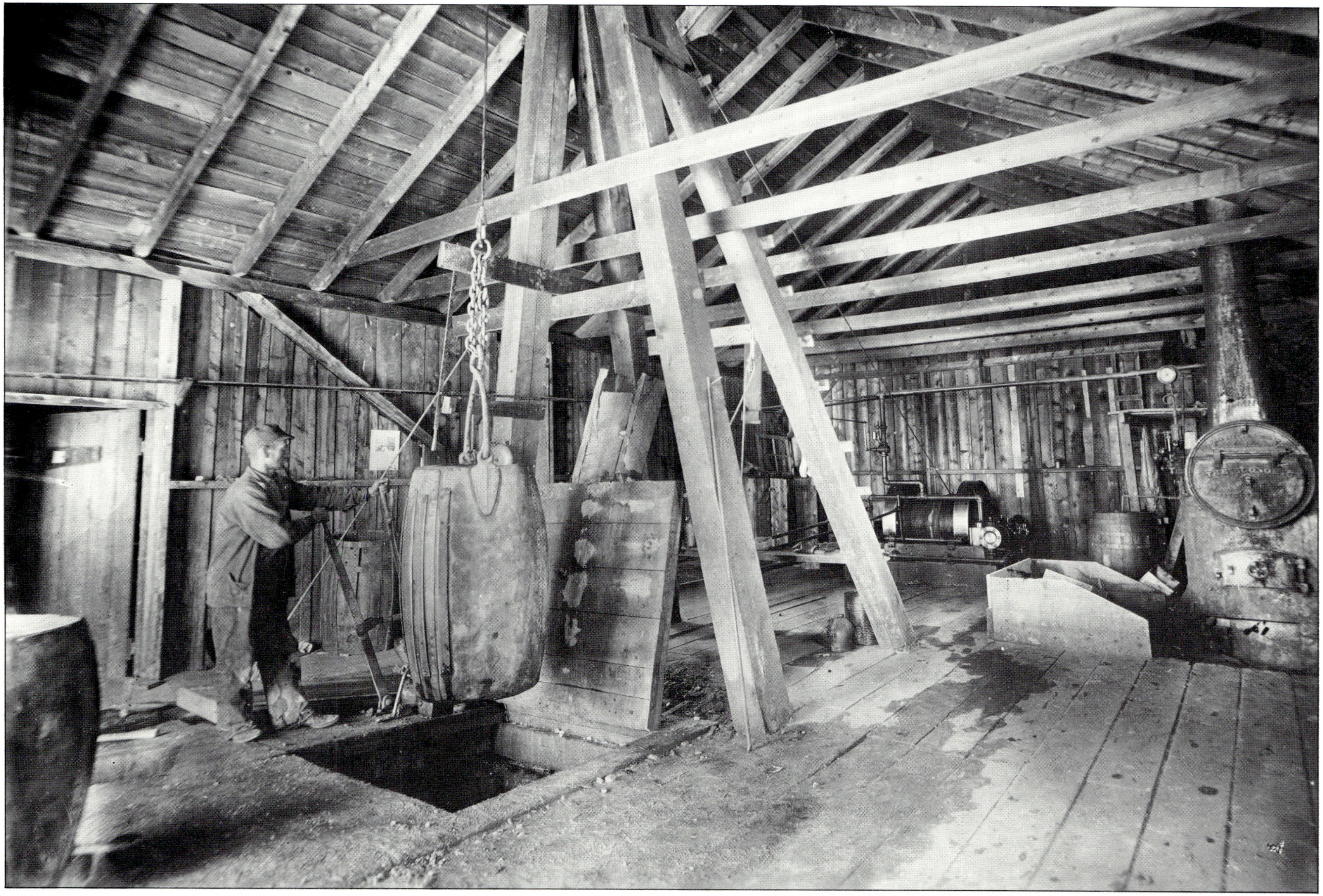

24. The interior of the shaft house shows the hoist and boiler. In this small mine the engineer operated the equipment from an unusual spot. He had to be careful to know where the bucket was at all times to avoid suddenly dropping it to the sump [bottom of the shaft], or literally raising the roof.

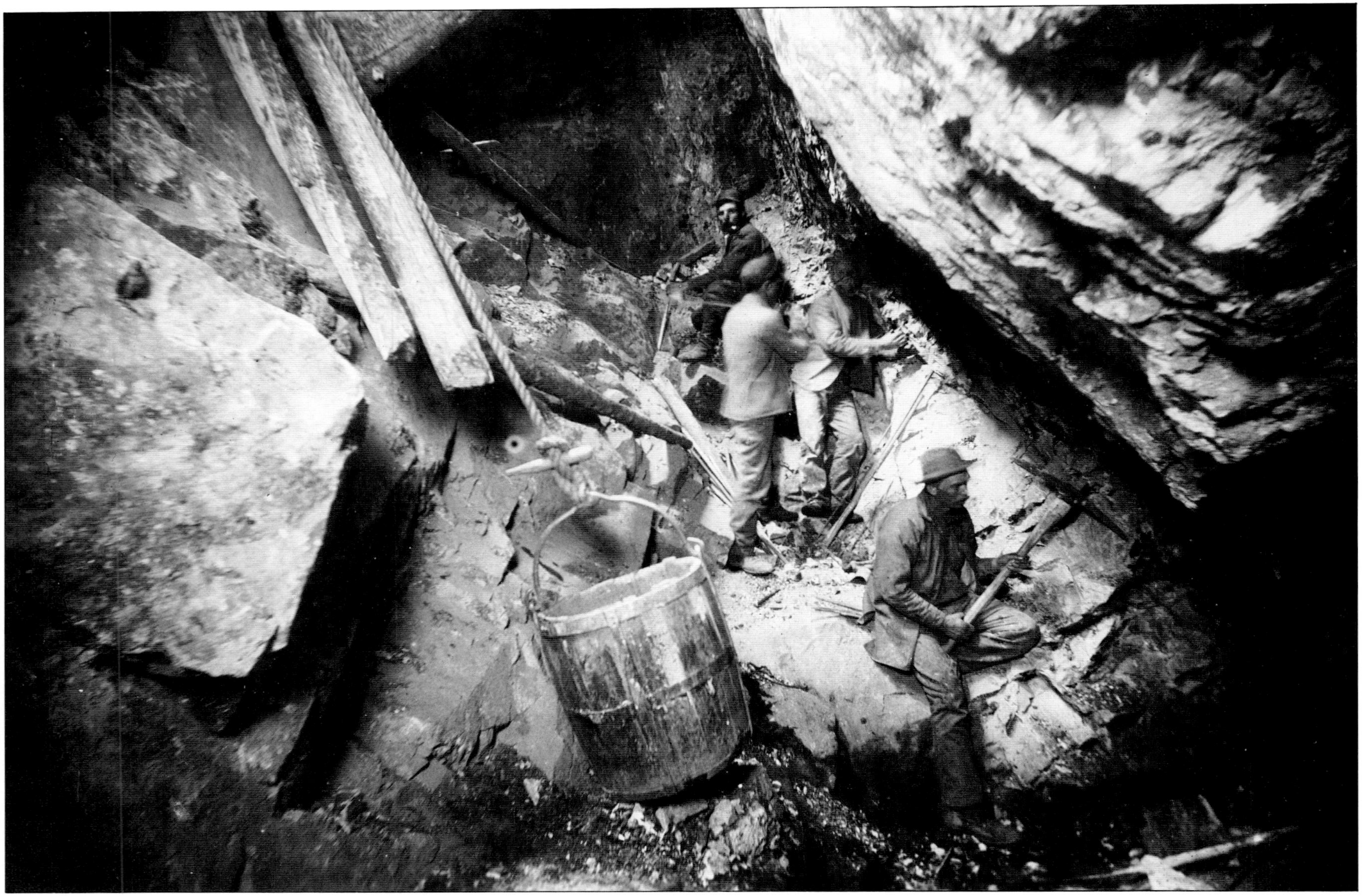

25. On going underground, the visitor quickly perceived that he had entered another world. This crew worked in what appears to be a stope, only recently opened. This is, perhaps, one of McLean's early photographs, because wooden buckets were soon replaced by metal ones. The rope hauling the bucket up a semi-inclined shaft indicates the surface cannot be too far away.

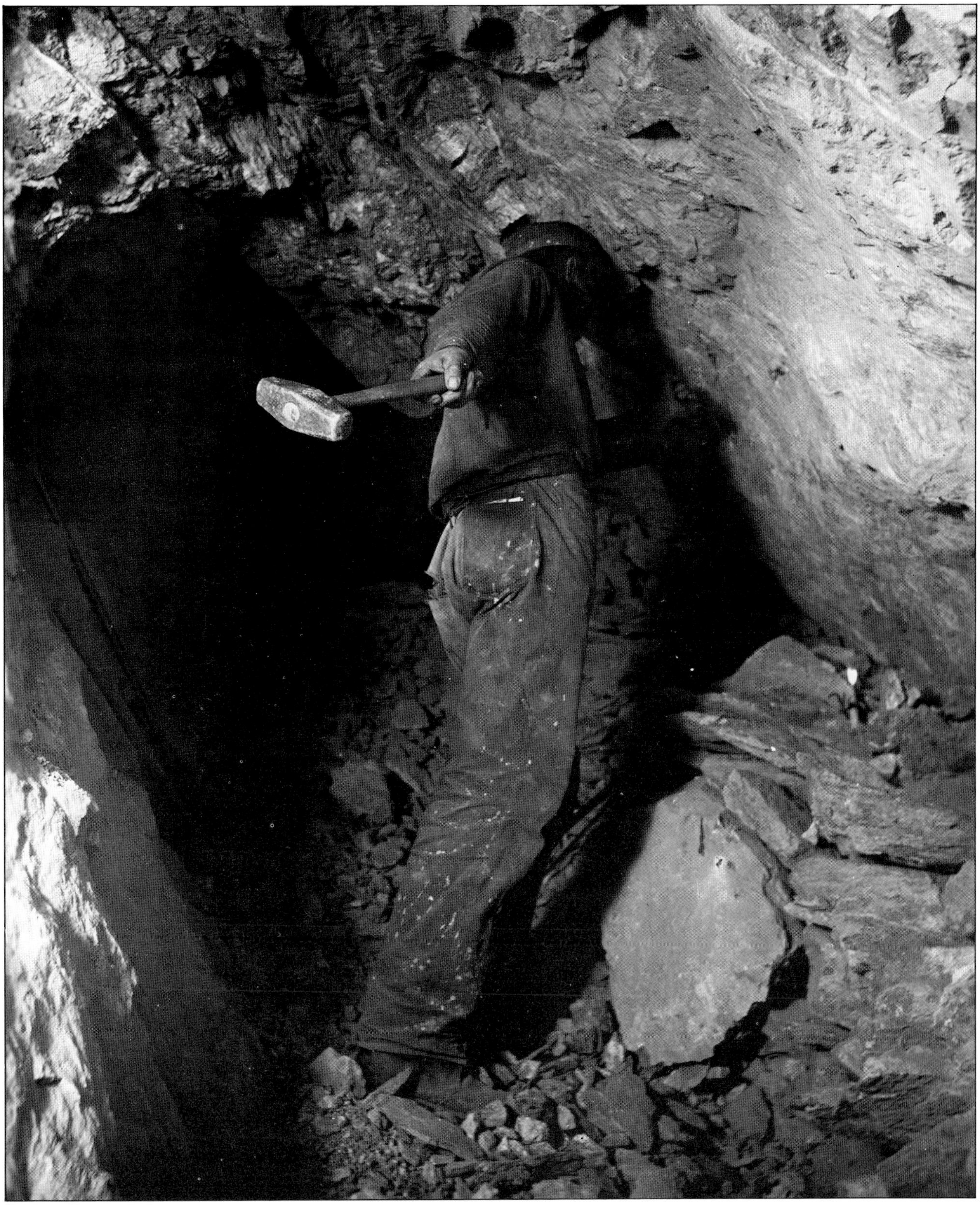

26. A single jacking miner holding and turning the steel with his left hand, swinging the hammer with his right. The hammer usually weighed four pounds, occasionally less. After a pattern of drill holes was completed, explosives were tamped into the holes, the fuse was set and lit, and the shift was over. The next crew mucked and trammed the ore and waste.

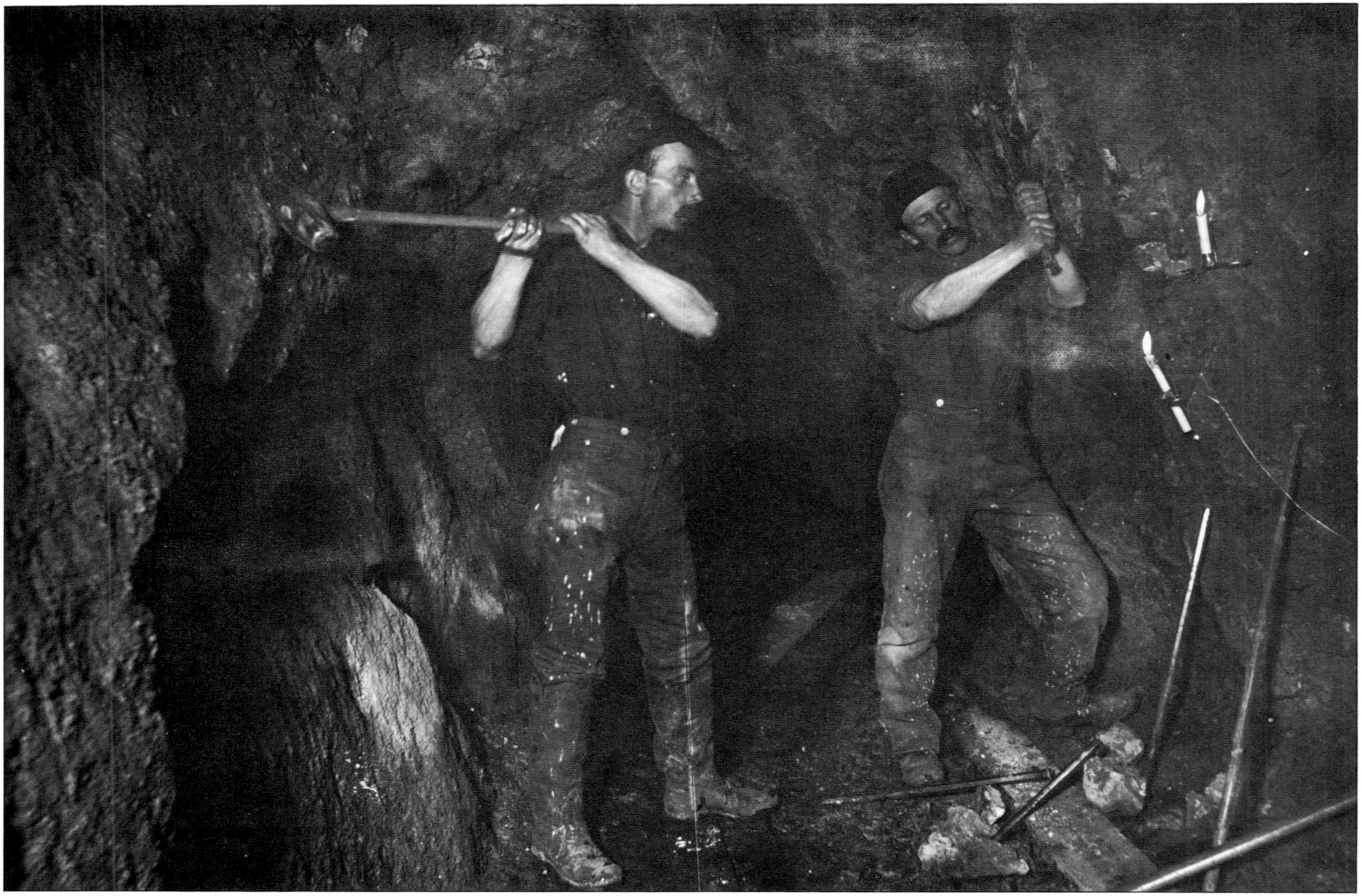

27. Traditional double jacking—one miner holds the steel bit, the other swings the hammer of eight to ten pounds. The steels, or drills, came in sets whose tip flares grew progressively thinner and whose length increased as the hole went deeper. Note those at the bottom. The "starter" was about a foot long, nearly $1\frac{1}{4}''$ in diameter.

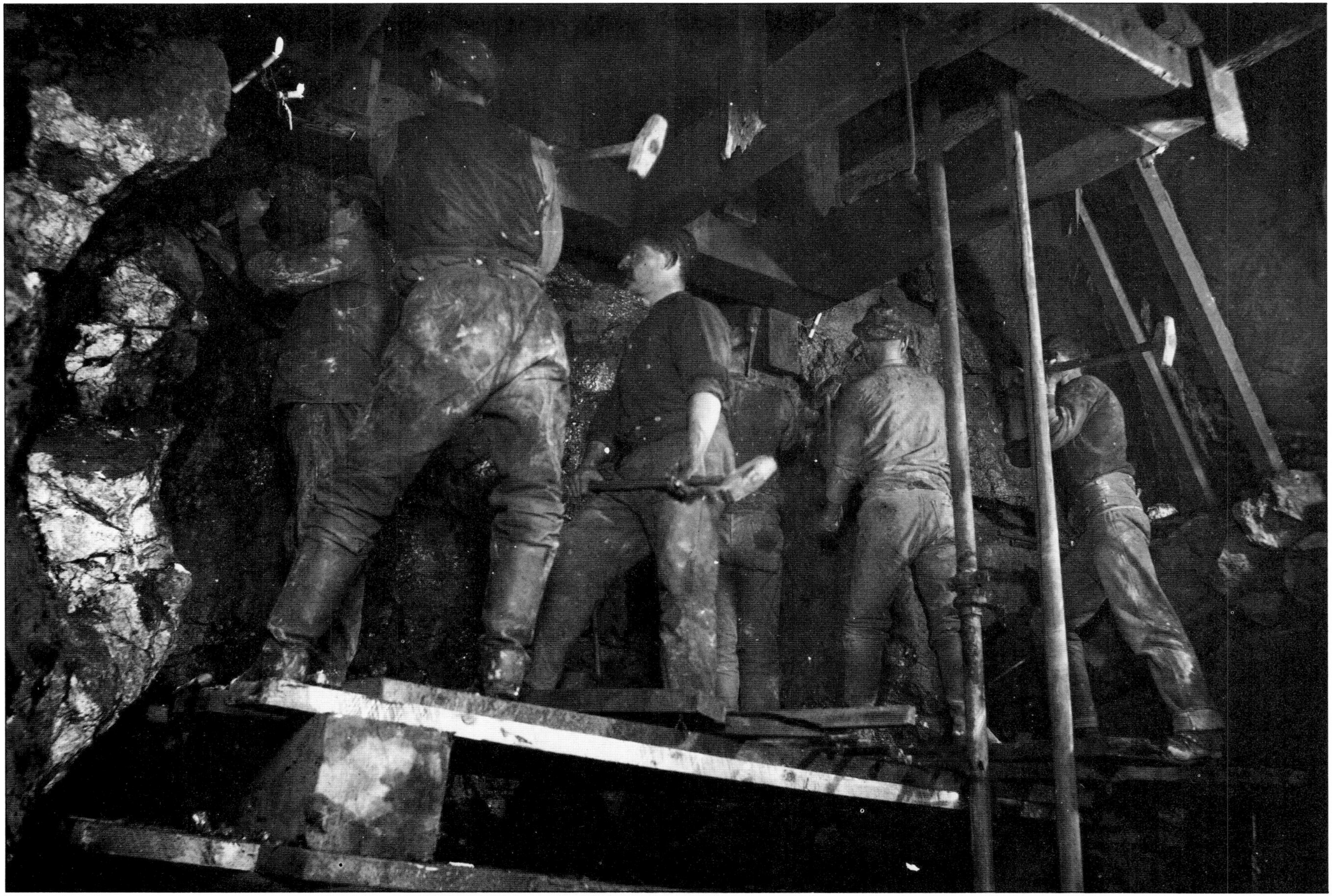

28. A more unusual three-man team, with two men pounding in disciplined, rhythmic succession. None of these miners could have worked a shift if it had not been for the blacksmiths who resharpened and retempered the drills; the steels wore out incredibly fast against hard rock.

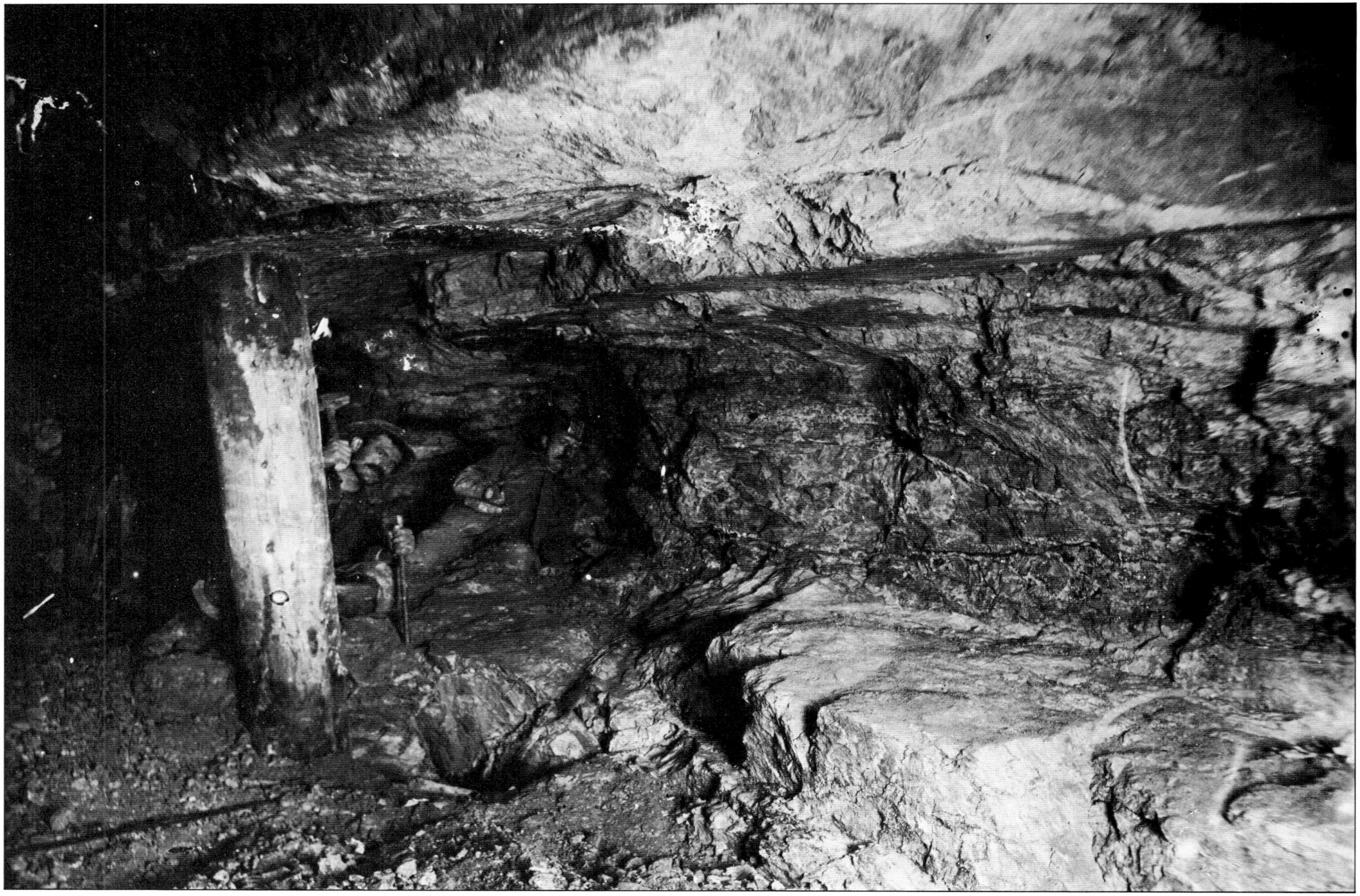

29. No one has ever said mining was easy, but certainly it was even more difficult when drilling on one's side. Generally the miners were supplied two or three candles whose combined burning time was intended to equal the length of the shift. If candles ran out and none could be found or borrowed, it meant a trip out to get more, or calling it quits for the day.

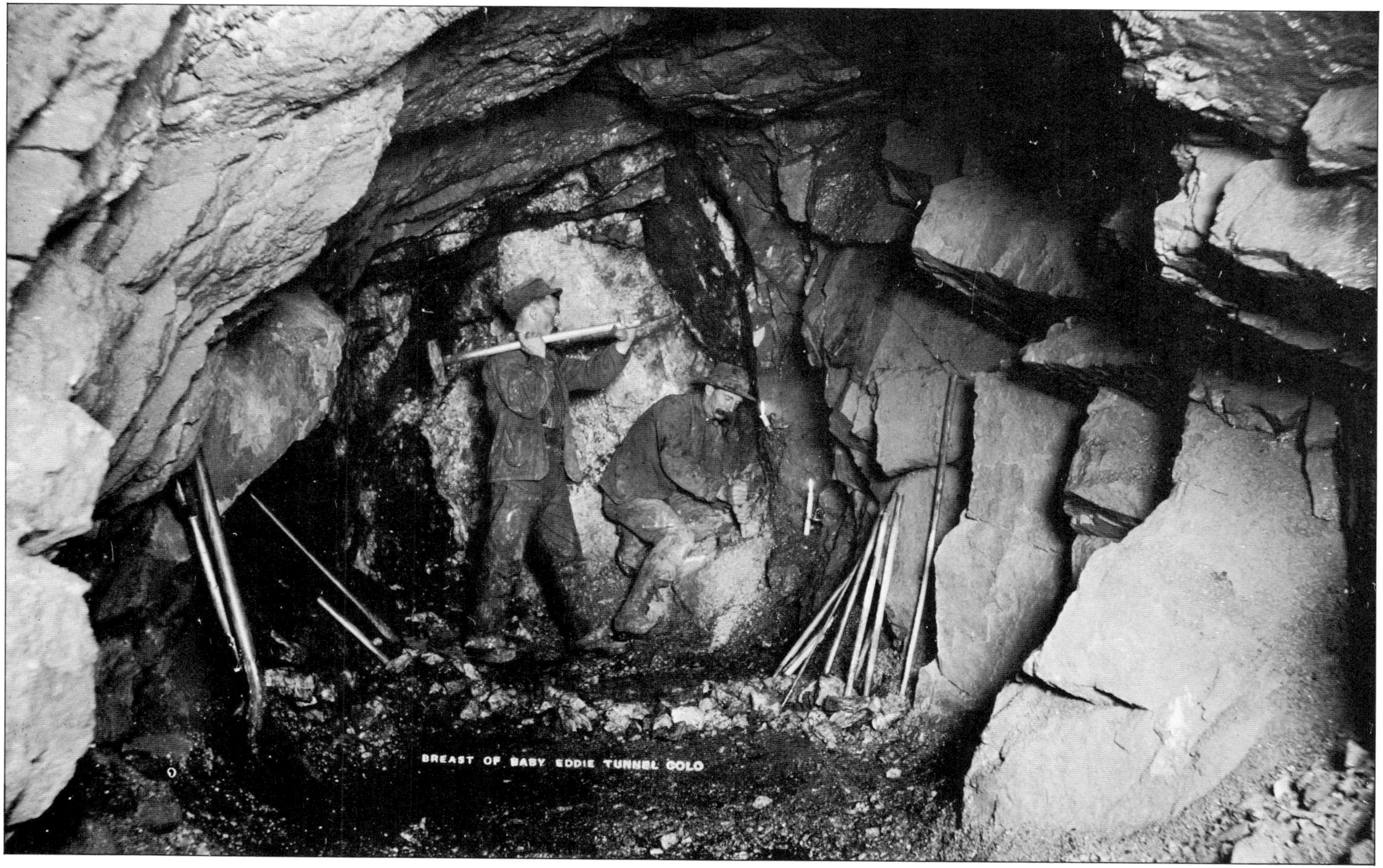

30. The breast, or working face, of the Baby Eddie Tunnel, seen earlier on the outside. Swinging a sledge in the semi-darkness of candlelight was an art in itself and carried a risk for the man holding the drill. After each blow, the steel had to be lifted slightly to clear the rock dust from the hole and then rotated to give the bit a new surface in which to cut. A story is told about a pair of Gilpin County miners who worked so long together that if the drillman wanted to change bits or clean out the hole, he would merely place his thumb over the top of the steel and his partner would not finish the blow.

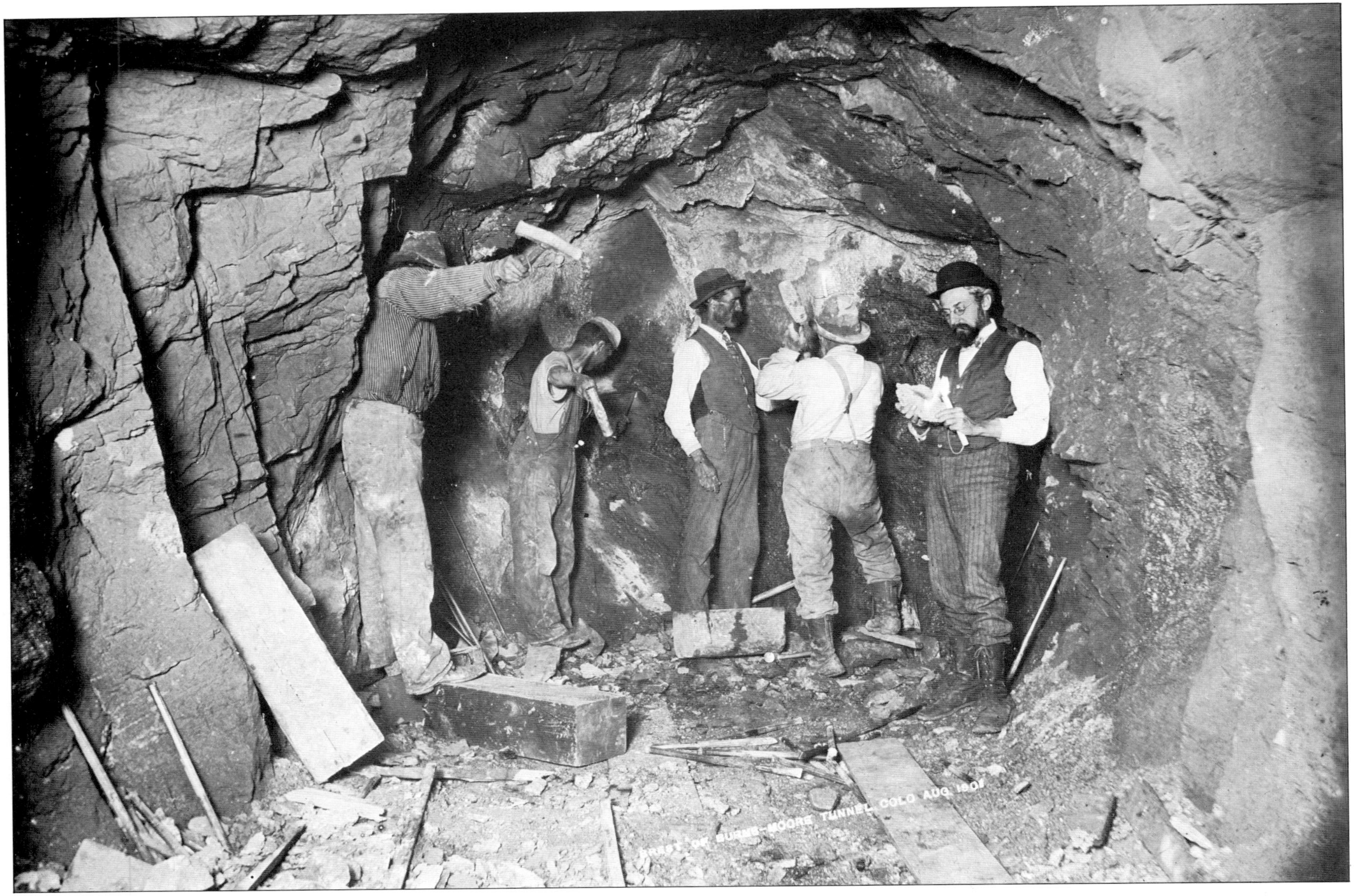

31. Two of these men were not part of the regular Burns-Moore Tunnel crew in August 1901. What they are doing there, other than studiously examining the ore sample and gingerly watching a miner drill, is a mystery. With all his skills, McLean could not capture the noise, dust, and smell of mining.

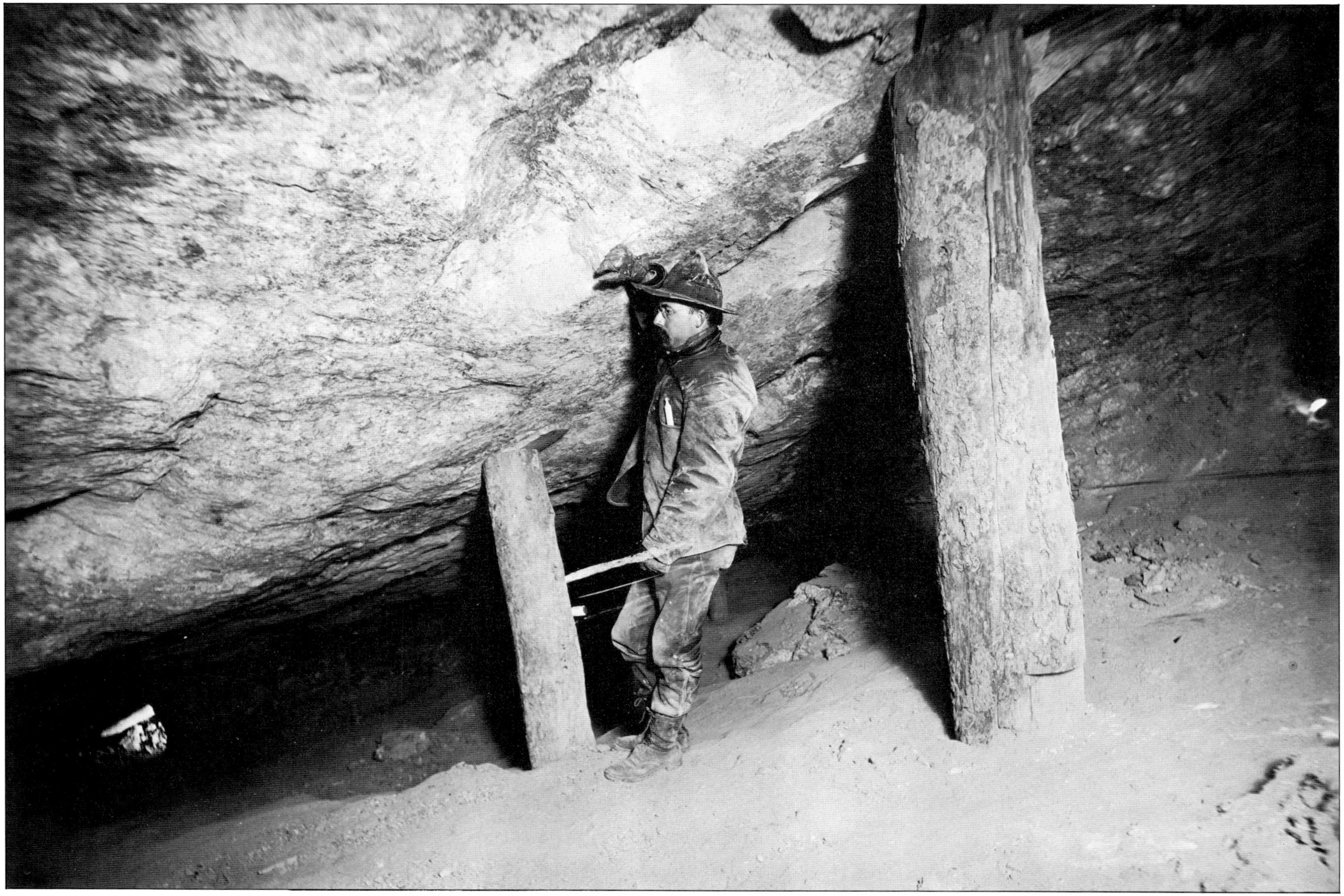

32. A mining engineer, or perhaps a shift boss, checks over developments. The light at the end of the tunnel, so to speak, indicates he was near the surface. Trained mining engineers were generally better accepted by the turn of the century, a progressive development for Clear Creek mining.

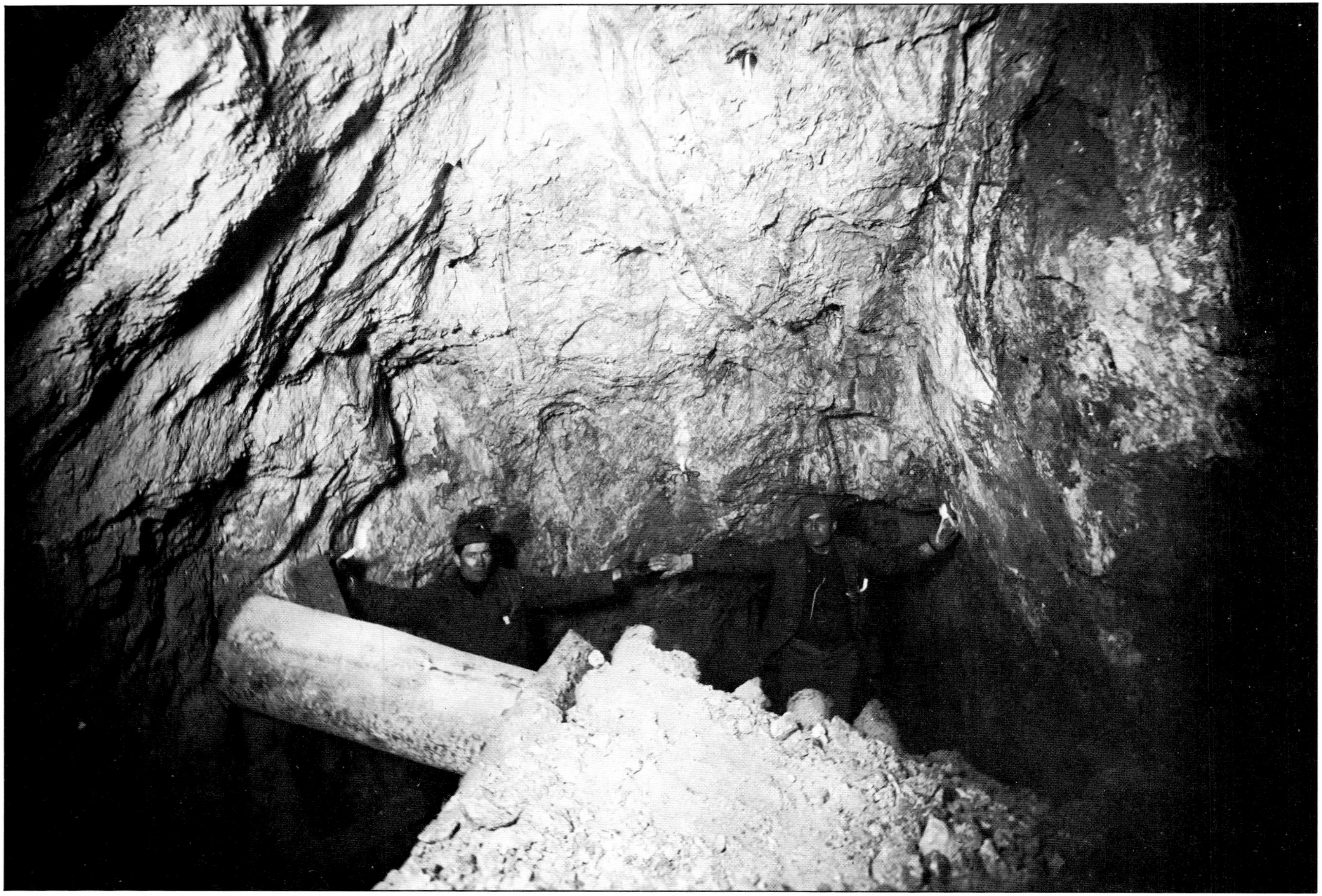

33. If the ore was as high in grade as these two miners indicate the vein was wide, then they had a bonanza mine. The miner's candleholder, shown above their hands, has become a valued antique today.

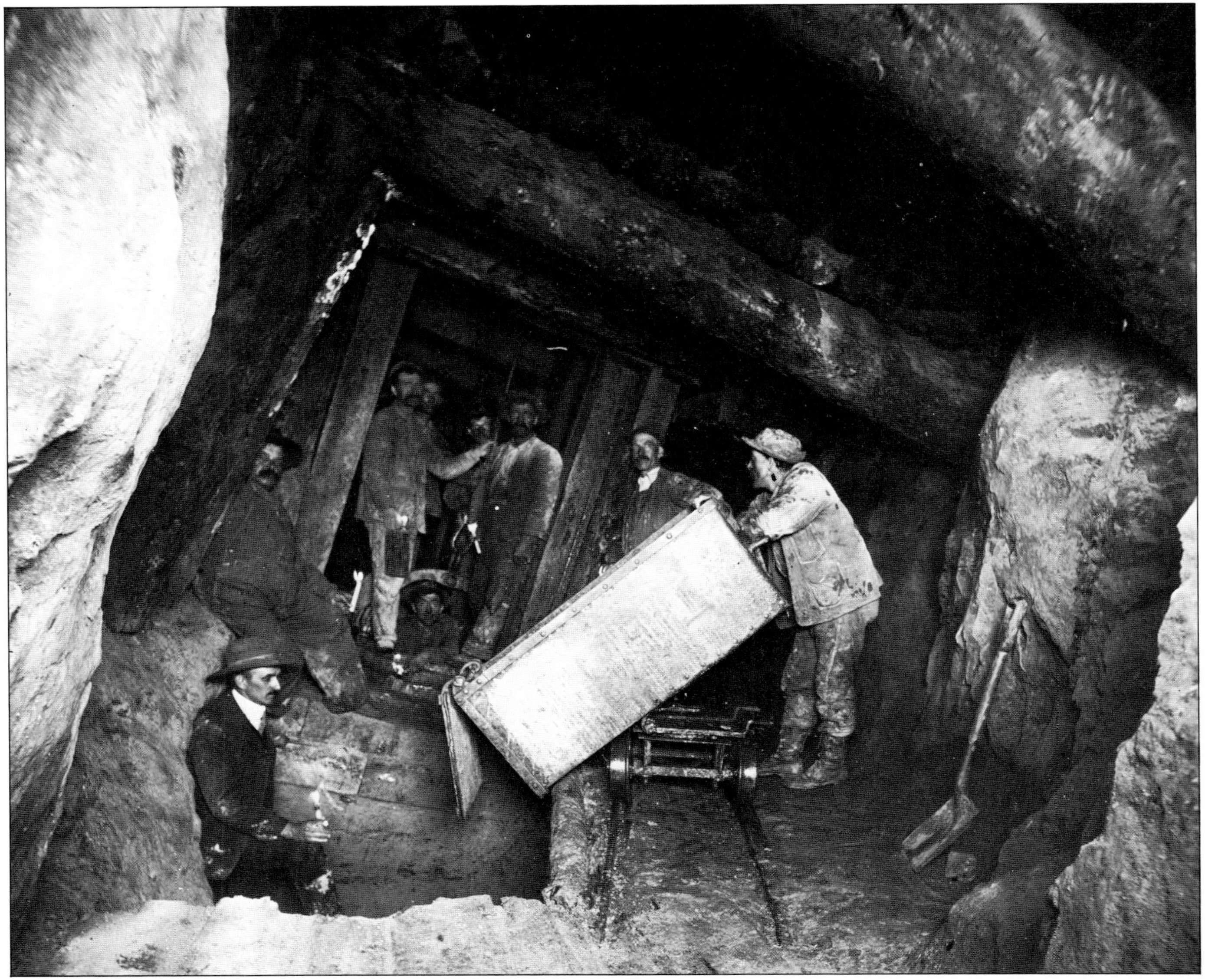

34. Normally no one would stand in an ore chute, but if you were the boss, it must have been permissible. The common one-ton ore car dumped endwise and also swiveled, and could be dumped sideways; it ran on rails of eighteen-inch gauge. In the background a group of miners stands on the edge of a shaft, and one rests in the bucket. Note the various types of timbering used in this mine, which must have been one of the larger ones.

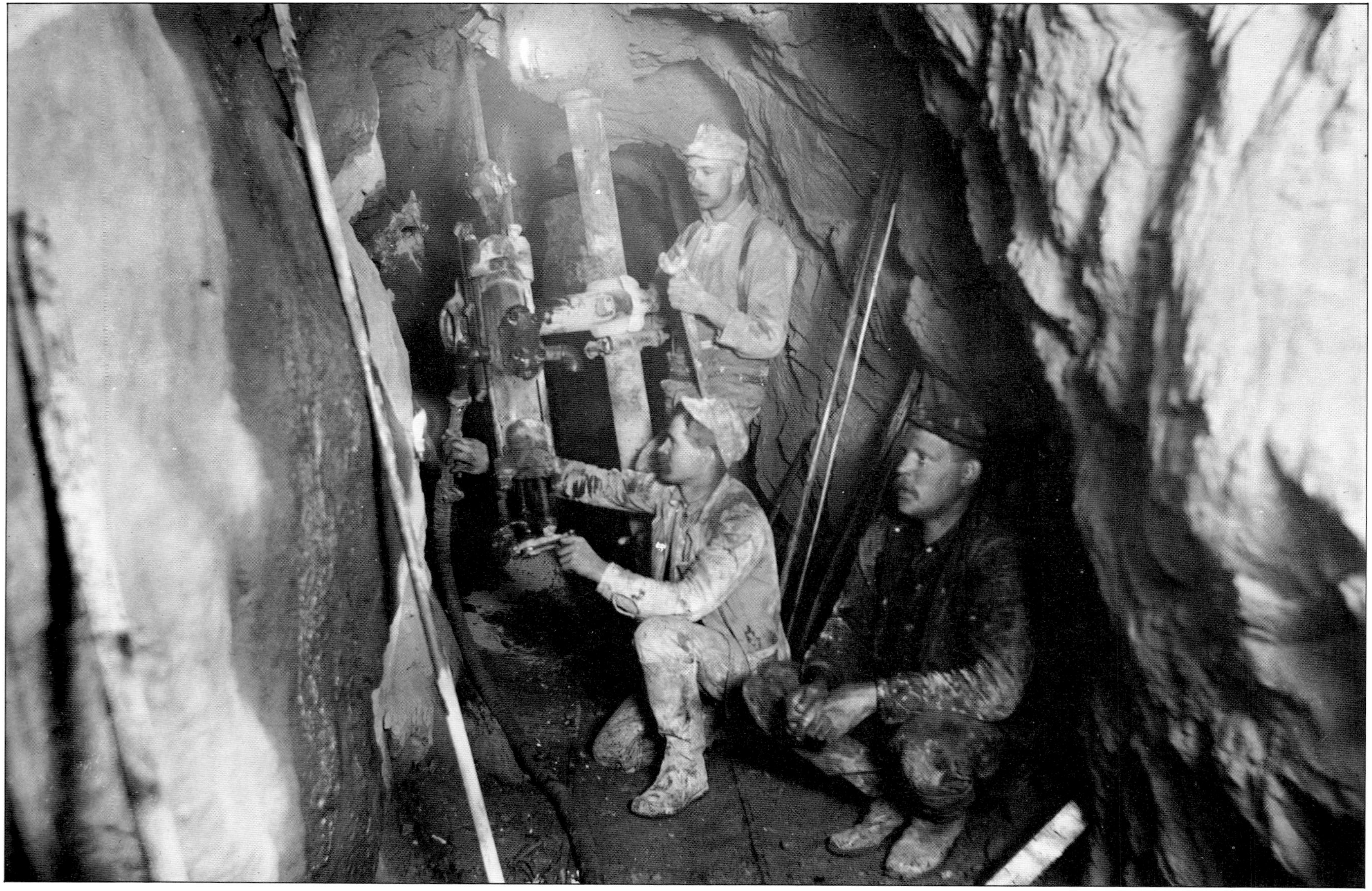

35. Poor devils; one wonders if this crew of young miners lived long enough to enjoy the rewards of their labor. Breathing the rock dust eventually resulted in silicosis, or miners' consumption. This is one of the early "widow makers," so indicated by the single hose. Not even the best double jack team could hope to compete with the machine drill, even if it did cripple or kill the drillers. The introduction of a second hose and hollow drill brought water into the hole and cut the dust.

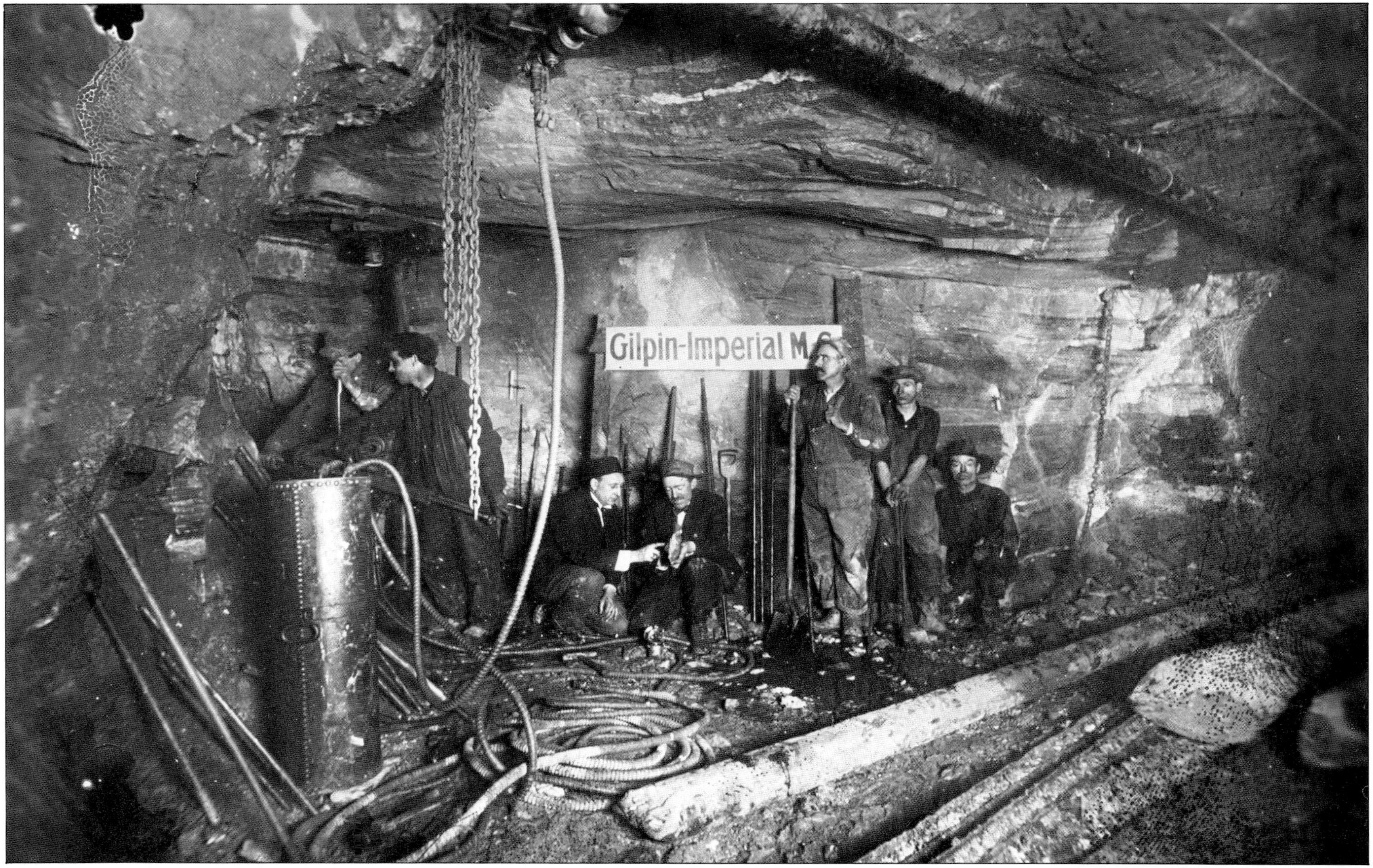

36. The Gilpin-Imperial Mining Company displays all the equipment needed for a successful mining operation; whether it had ore to match its expectations is not known. The compressed air drill quickly replaced the inefficient steam drill; Clear Creek County helped pioneer the former with the drilling of the Burleigh Tunnel in 1868-71 on Sherman Mountain above Georgetown. The jack tank at the left produced a steady air pressure and condensed any water in the air before it got into the drilling machine.

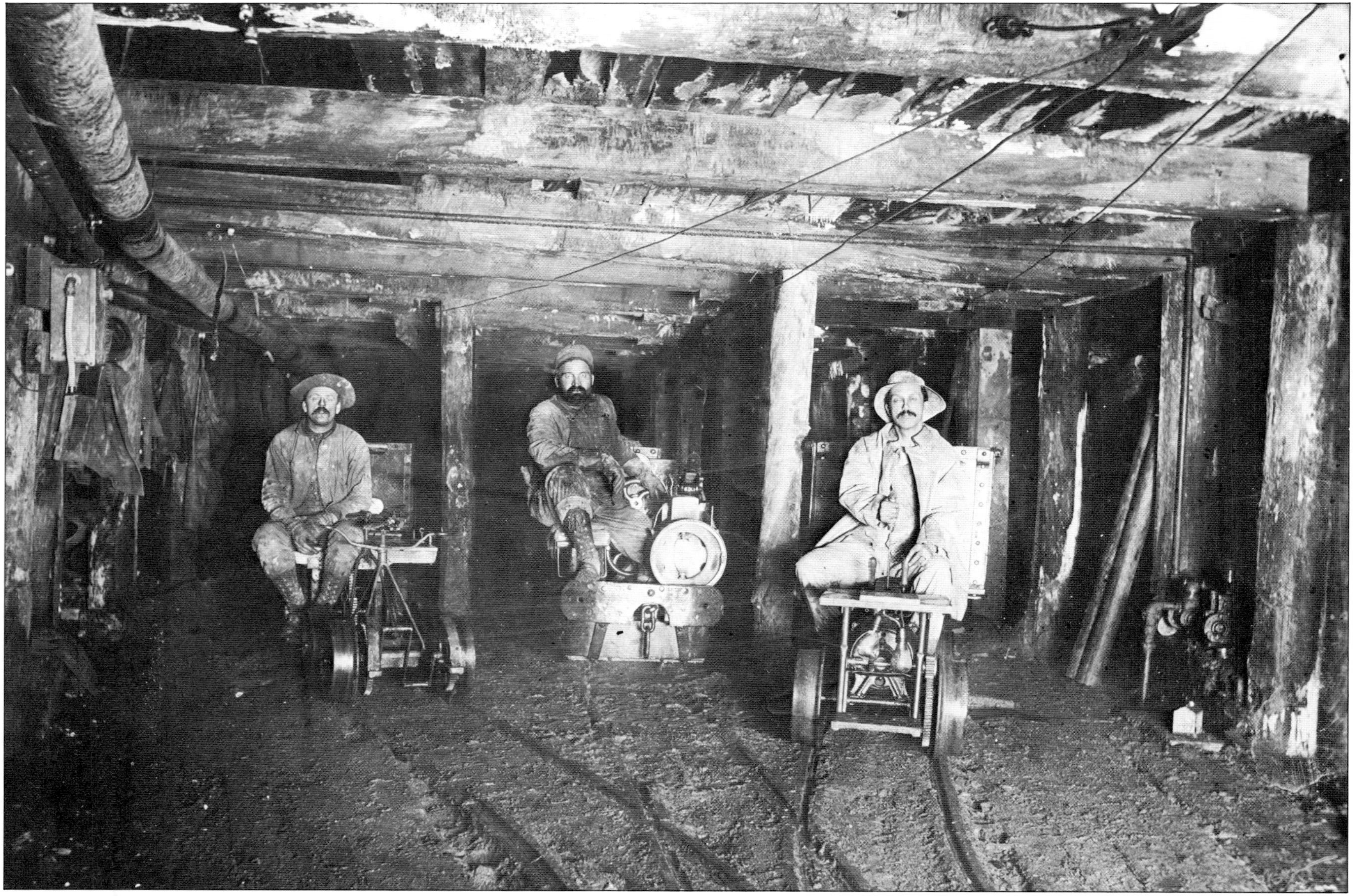

37. Tramming by hand or animal power was slow, hard work, even if the drifts and tunnels were so engineered as to have a slight downhill grade for the full car. Machine and electric engines replaced the older methods as soon as possible. The man on the left pedaled his tram; the other two operated similarly to the electric trolley. This may be the Argo Tunnel.

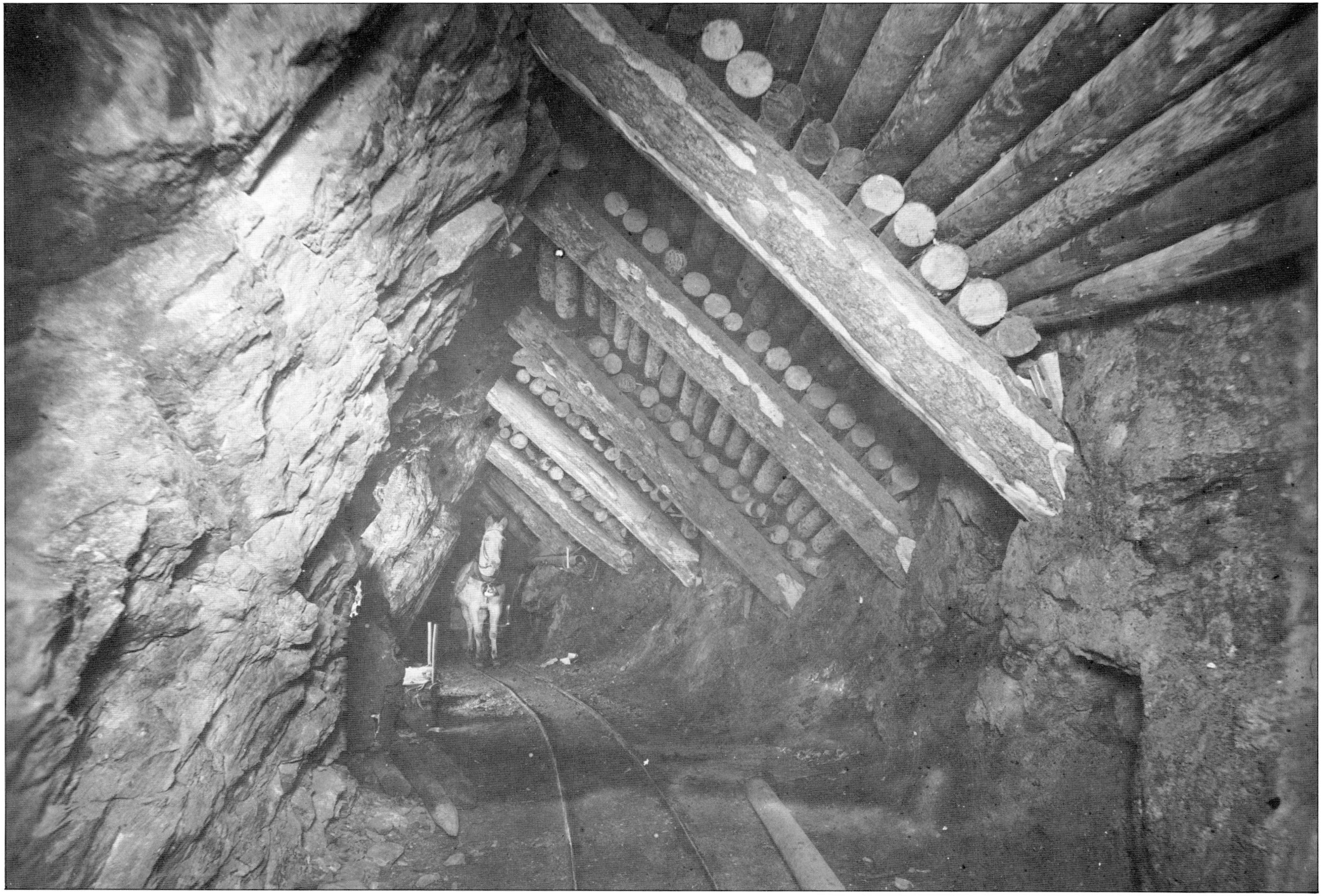

38. Few mines did not require timbering to support the roof or sides of a mine, to crib the shaft, and do a multitude of other things. This timber set supports part of the roof under which an ore train is about to pass. The hills in the Clear Creek Valley were denuded to supply this need.

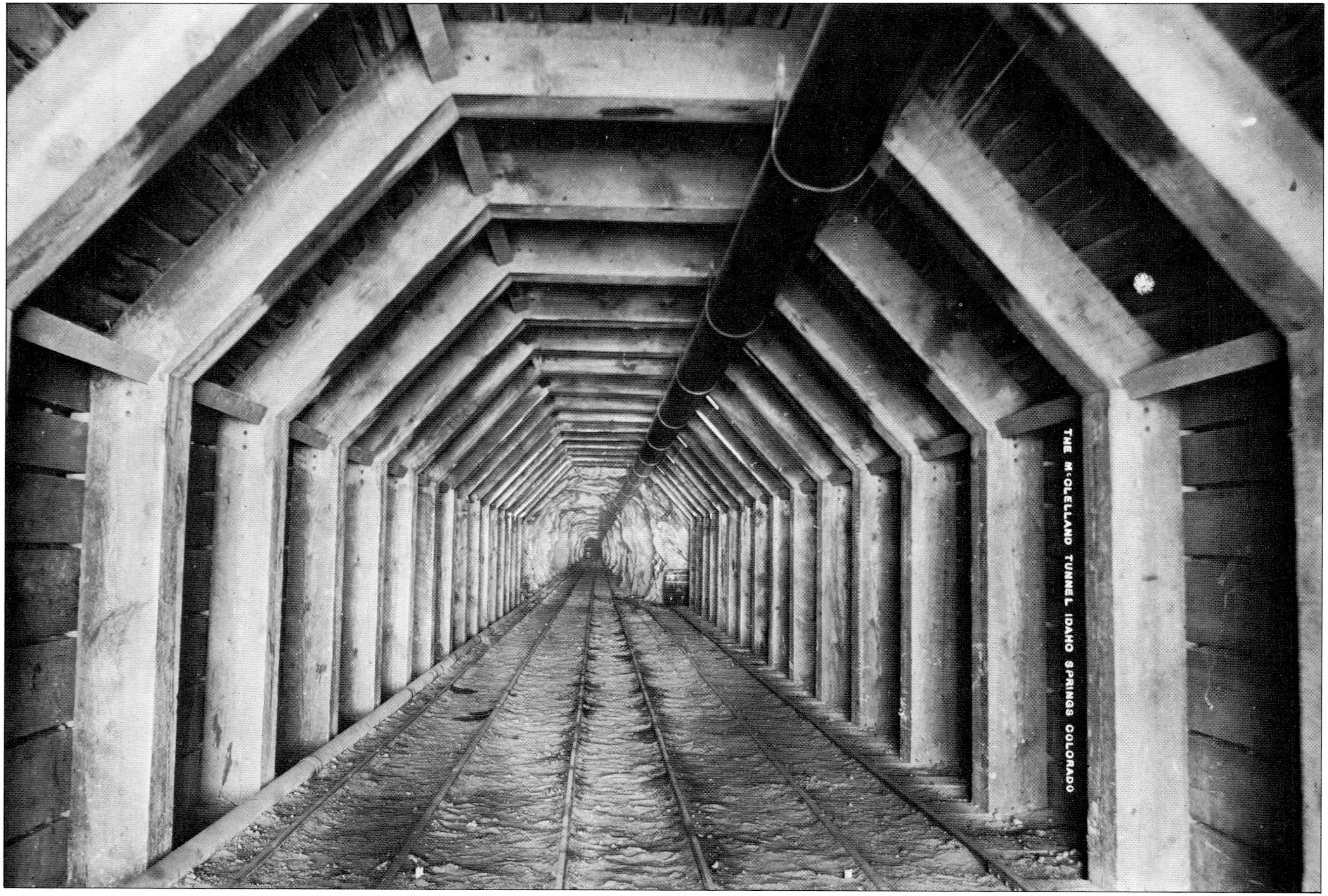

39. A classic example of tunnel construction, the McClelland Tunnel. Far down the tracks the light of an ore train can be seen. The theory of a tunnel was a fine idea, but Clear Creek mining had some bad experiences when theory met up with reality.

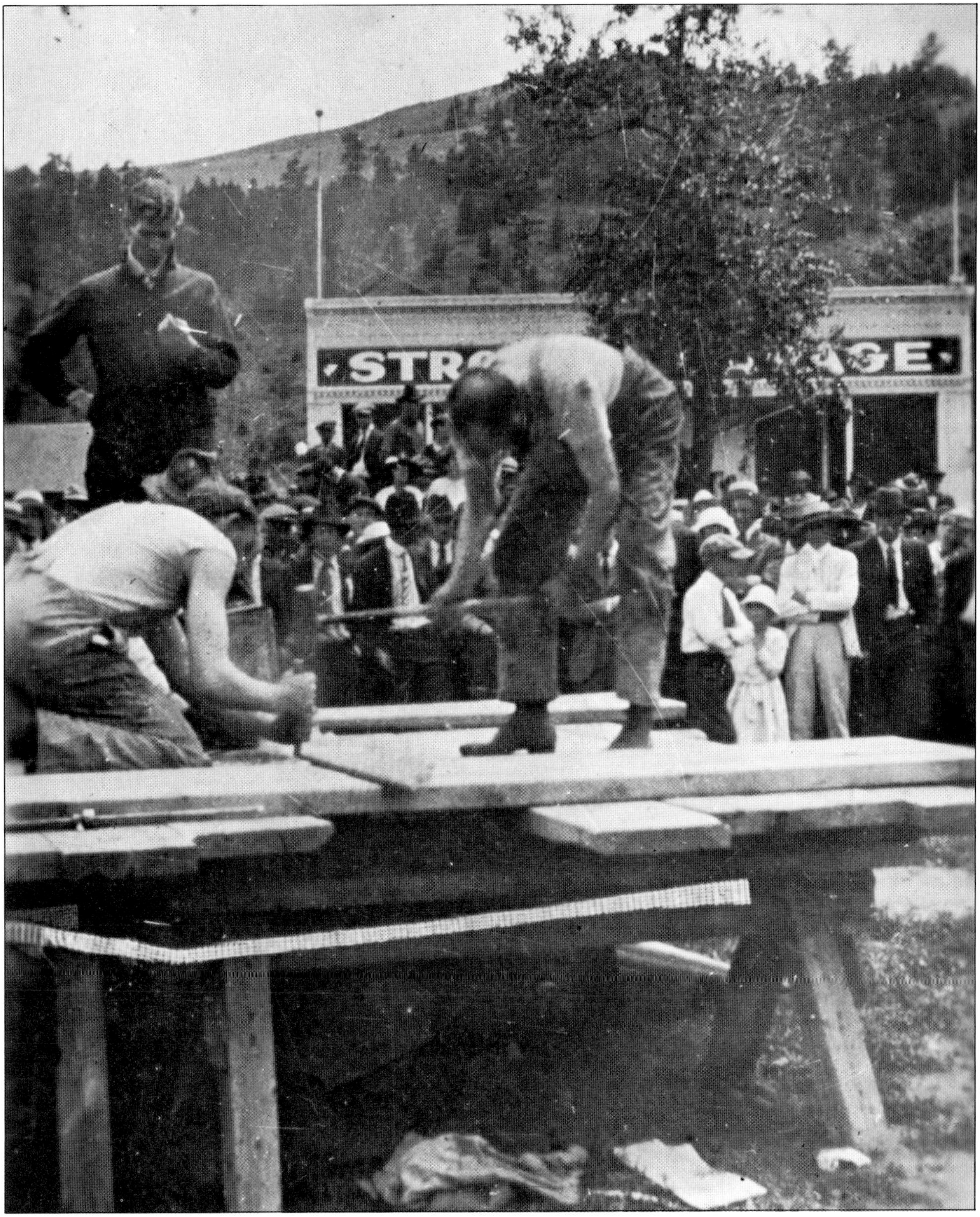

40. At the turn of the century and later, miners took their single and double jacking skills out-of-doors in contests that emphasized speed. By then the need for these skills was diminishing, but they still drilled against other miners, and against time, for the excitement, prize money, and glory involved. Betting heightened interest.

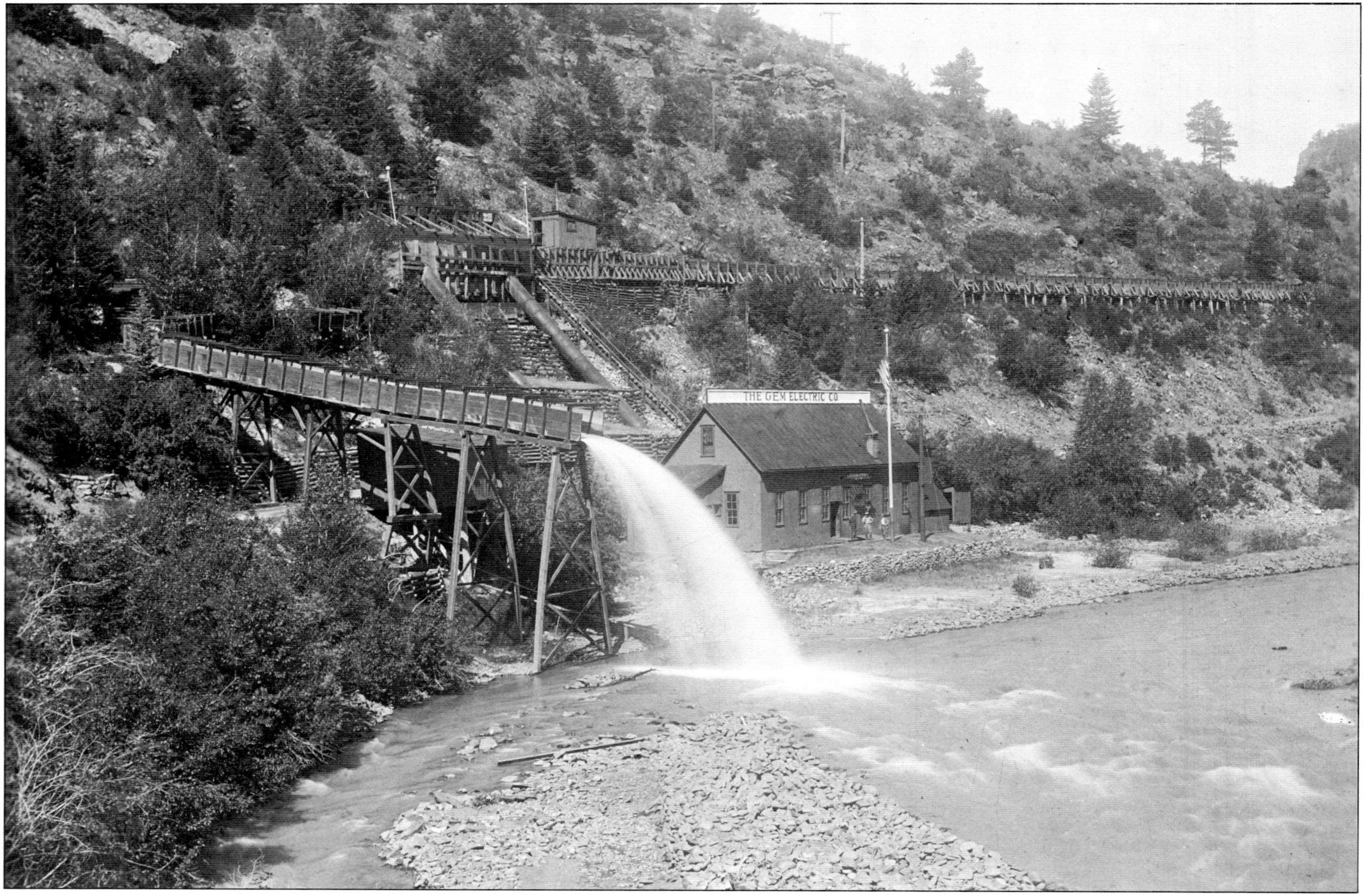

41. Electricity and railroads changed mining from the days of '59. The Gem Electric Company was one of the small generating plants in Clear Creek County. By 1908 electric power extended throughout much of the valley. The system of flumes shown here is reminiscent of the days of placering.

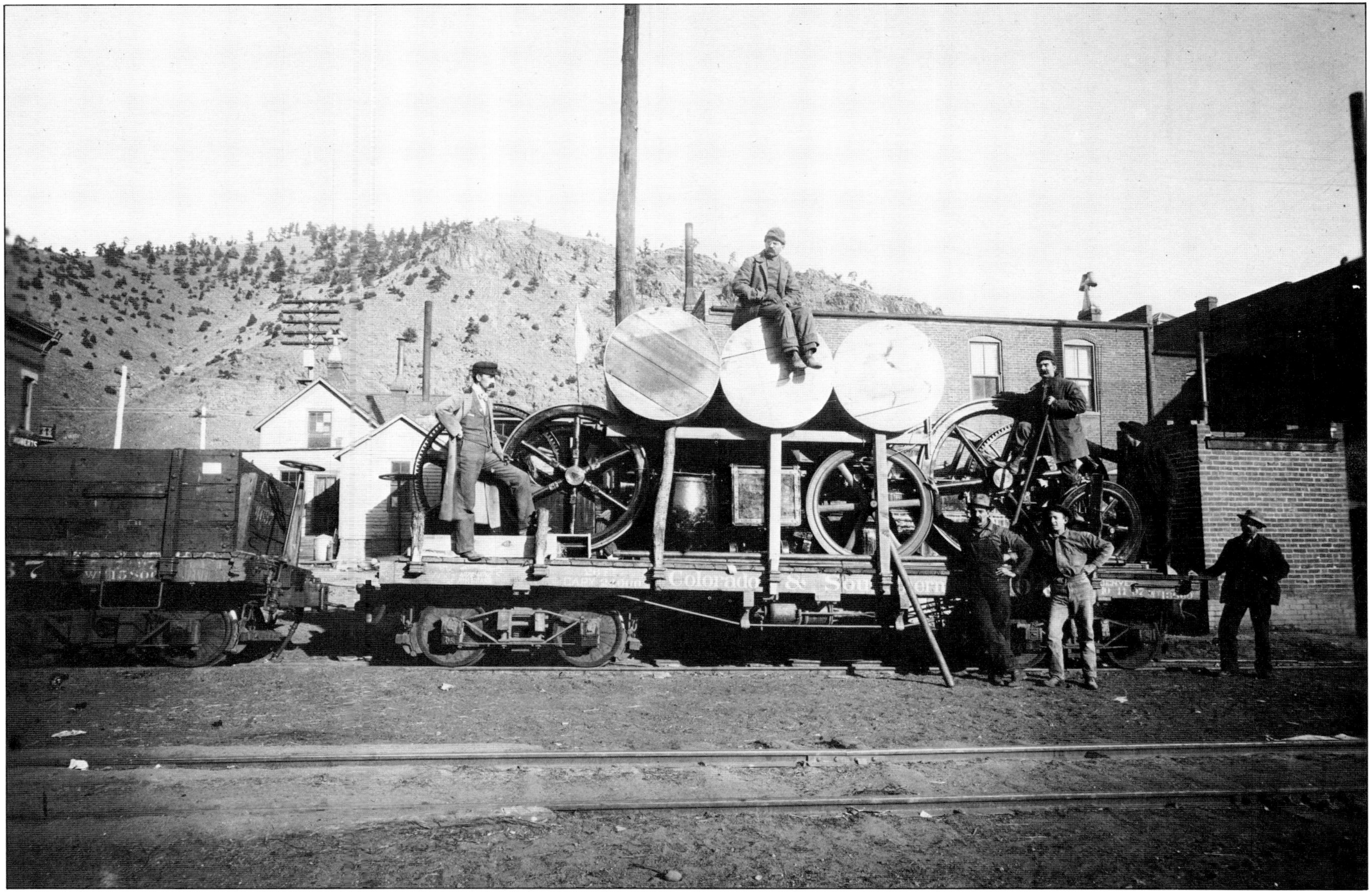

42. Imagine hauling this steam hoisting equipment by wagon to Idaho Springs or Georgetown; it was bad enough from station to mine. The pioneering Colorado Central Railroad eventually fell into the grasp of the larger Union Pacific and finally became part of the Colorado and Southern Railroad in 1899.

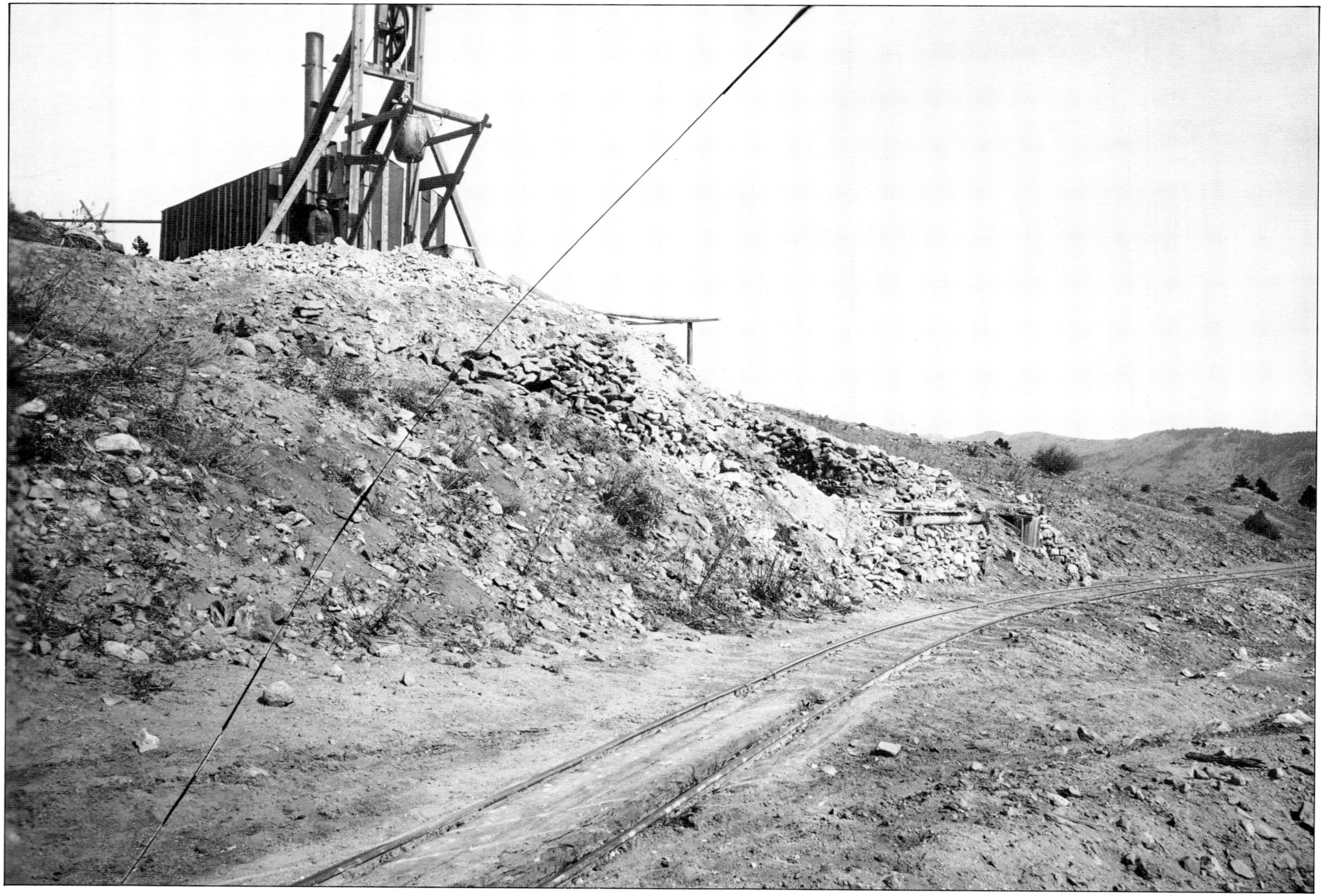

43. Every mine owner dreamed of having the railroad this close to his claim; few were so lucky. The mine itself was small, but notice the classic headframe, in some areas called a gallows frame, supporting the pulley for the hoisting rope.

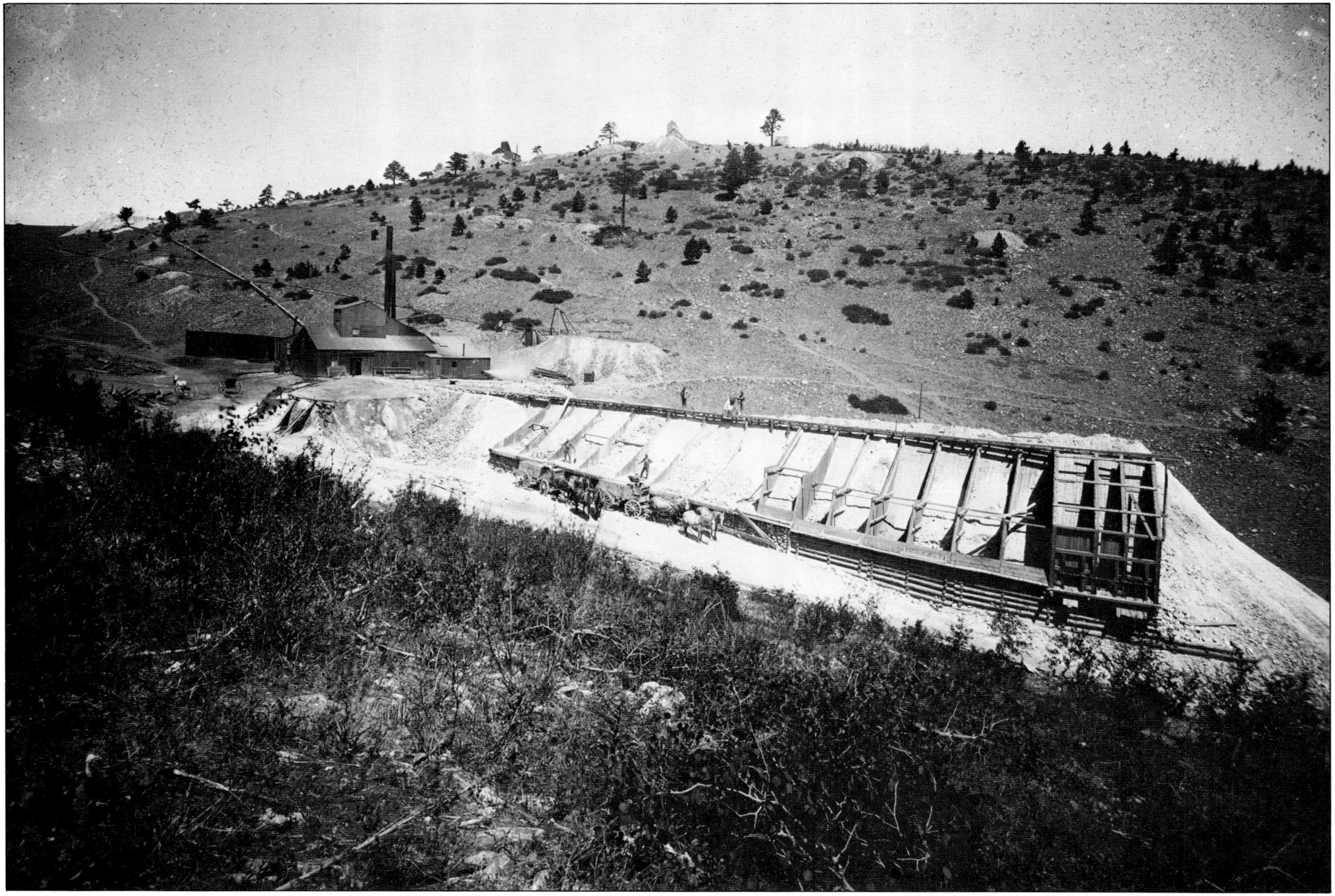

44. For those not fortunate to have the railroad nearby, hauling ore meant using wagons or, for the really isolated mines, mules or burros. These elaborate loading bins were at the Gem Mine. The carriage to the right of the main building may be McLean's.

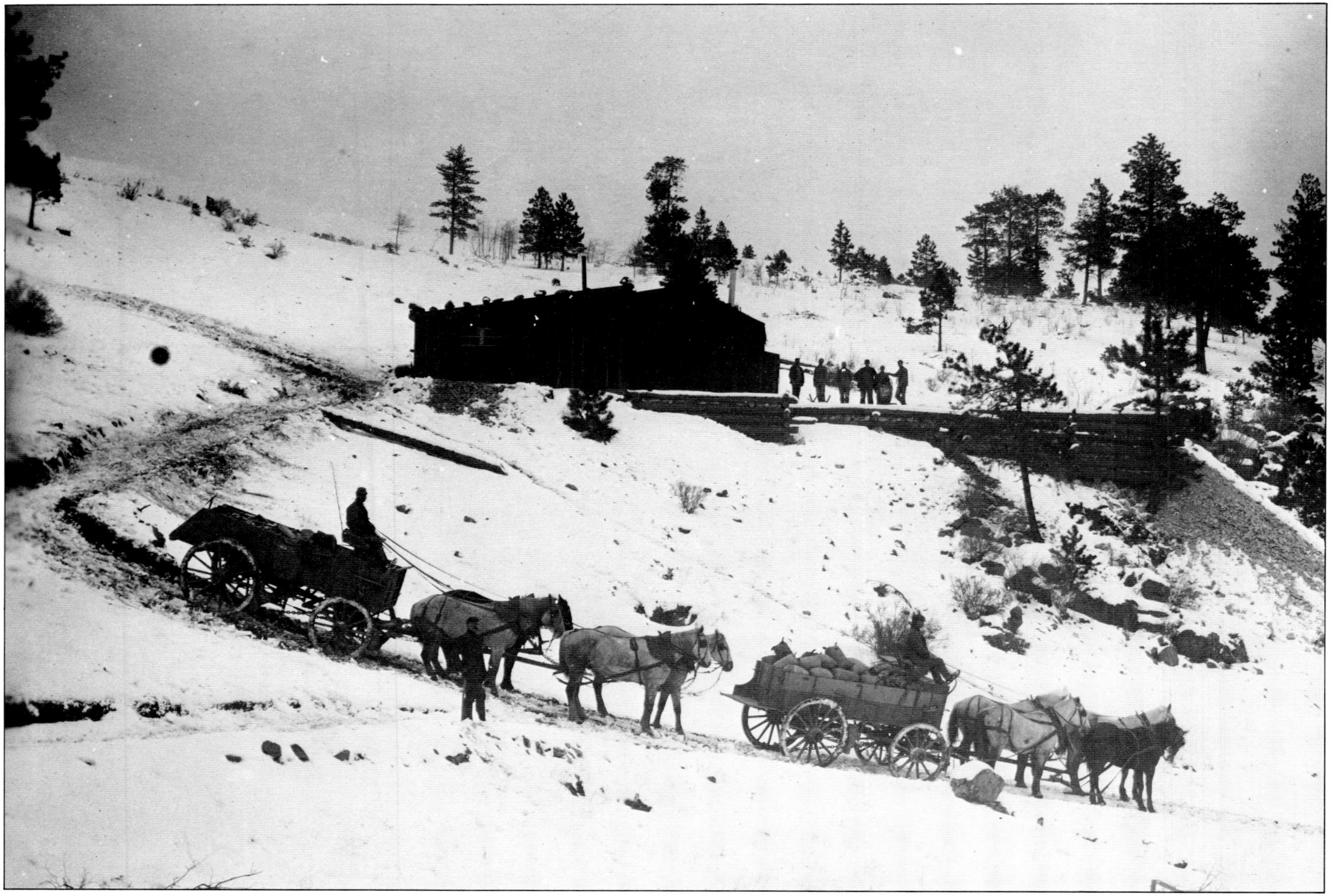

45. Winter freighting was perhaps the worst; the rear wheels of these wagons are chained. Many accidents occurred when the horses were unable to hold the load back. Steep, sharp curves were extremely hazardous to the freighters, who faced a long, cold day on one of these runs.

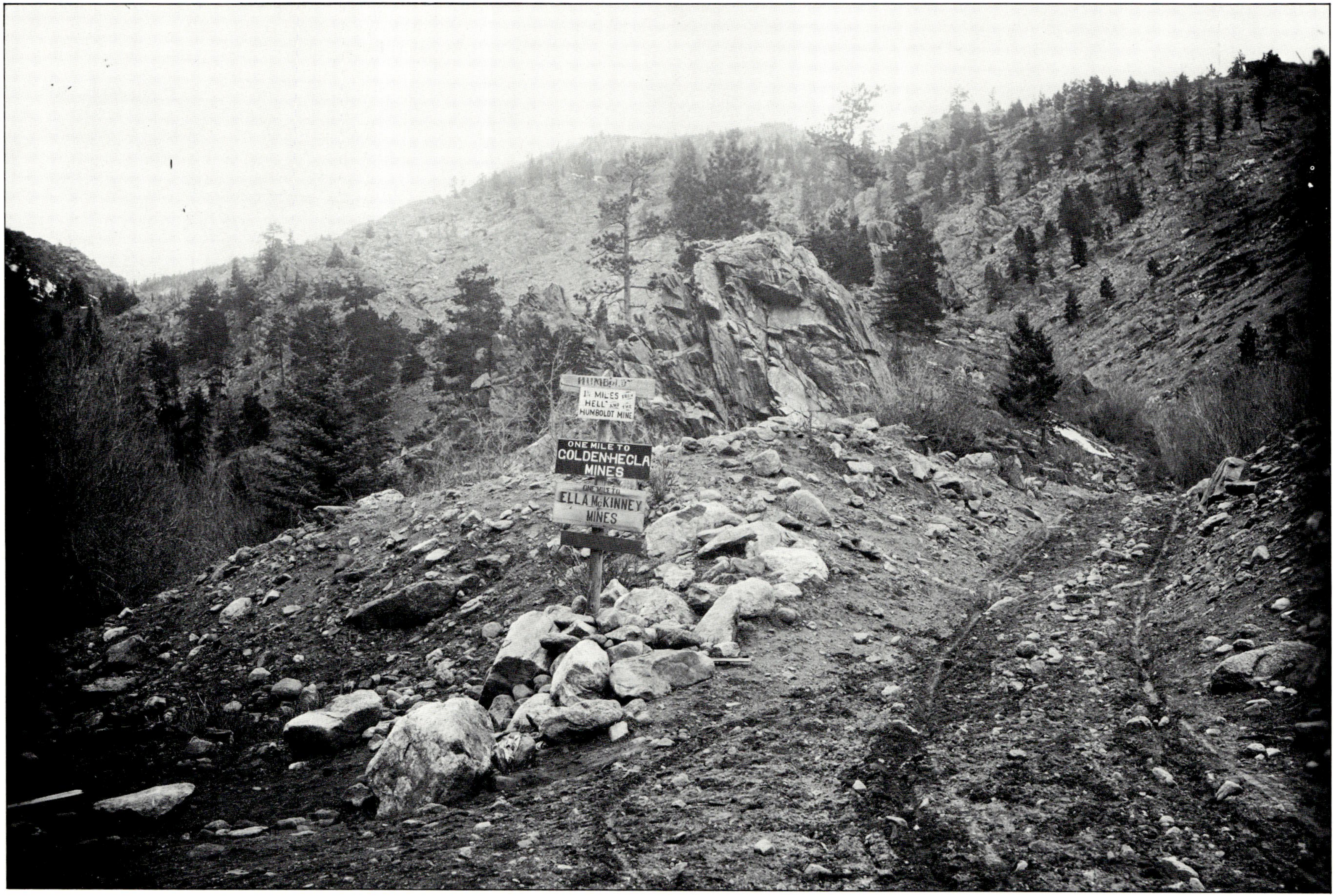

46. The roads might more appropriately be called trails, but over them rumbled the ore and supply wagons, as well as miners, owners, stockholders, and others going to and fro. On this spring day, mud would have been a problem.

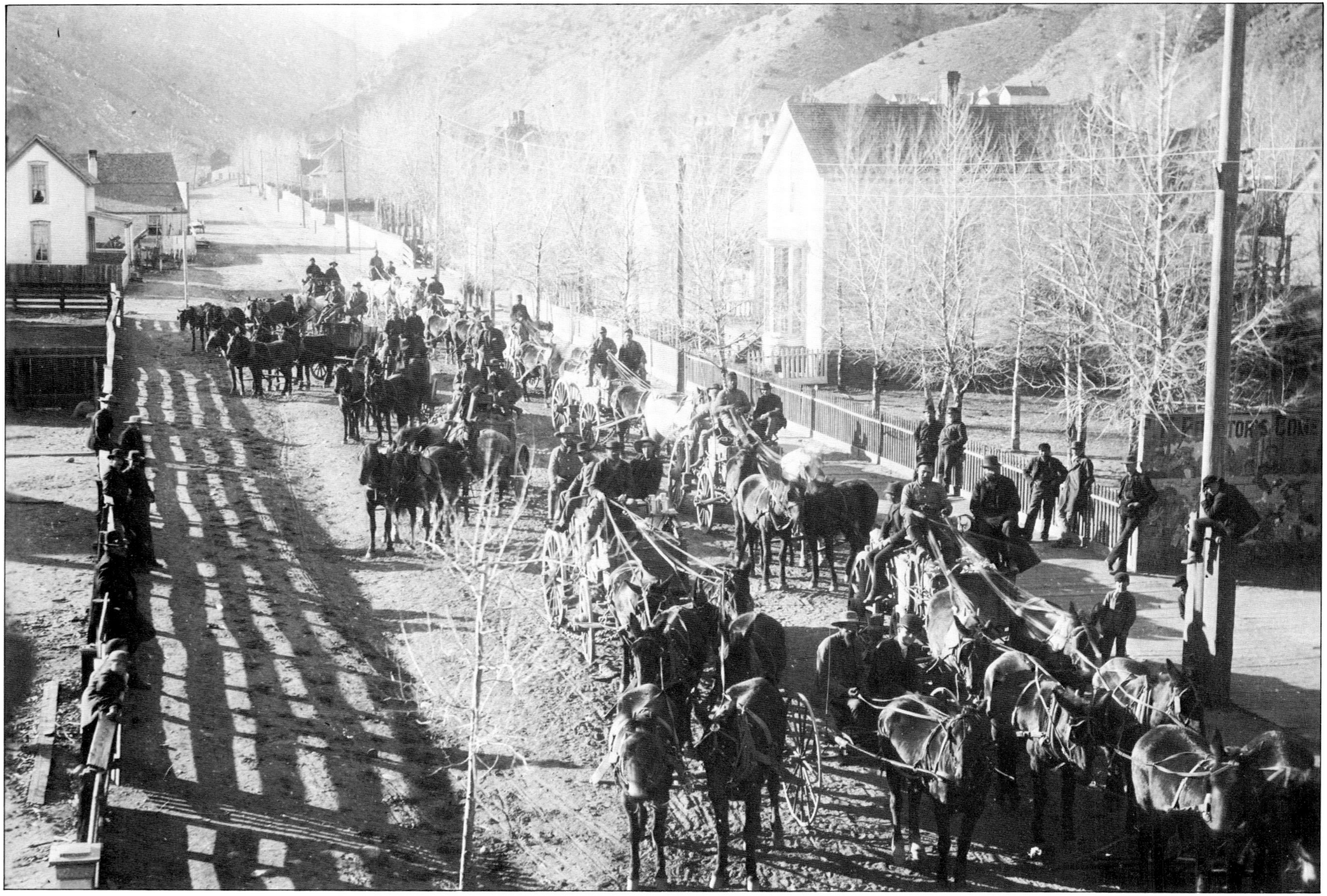

47. Lamartine Mine ore wagons in Idaho Springs—both four- and six-horse teams are stopped here. On steep grades, horses might push as well as pull, and it took expert teamsters to handle these heavy loads and navigate the mountain roads.

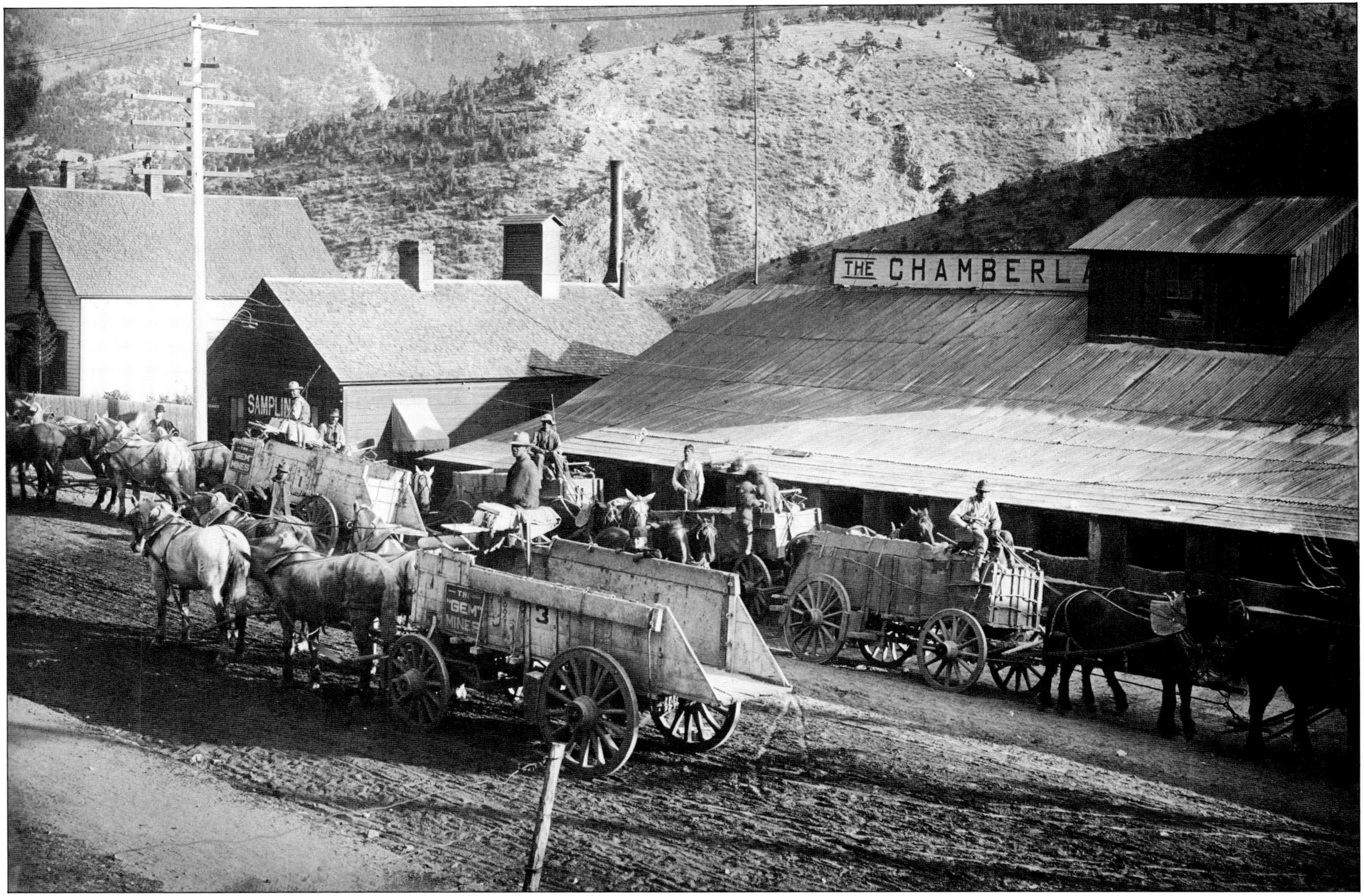

48. The goal of the freighter and miner was to get the ore to the mill or smelter. The Chamberlain Sampling Works at Idaho Springs thoroughly mixed the ore, took out a sample, and ran an assay. The ore would then be purchased on the basis of valuable metals found. Chamberlain either concentrated the ore by removing some of the gangue (waste rock) or sold it directly to a smelter.

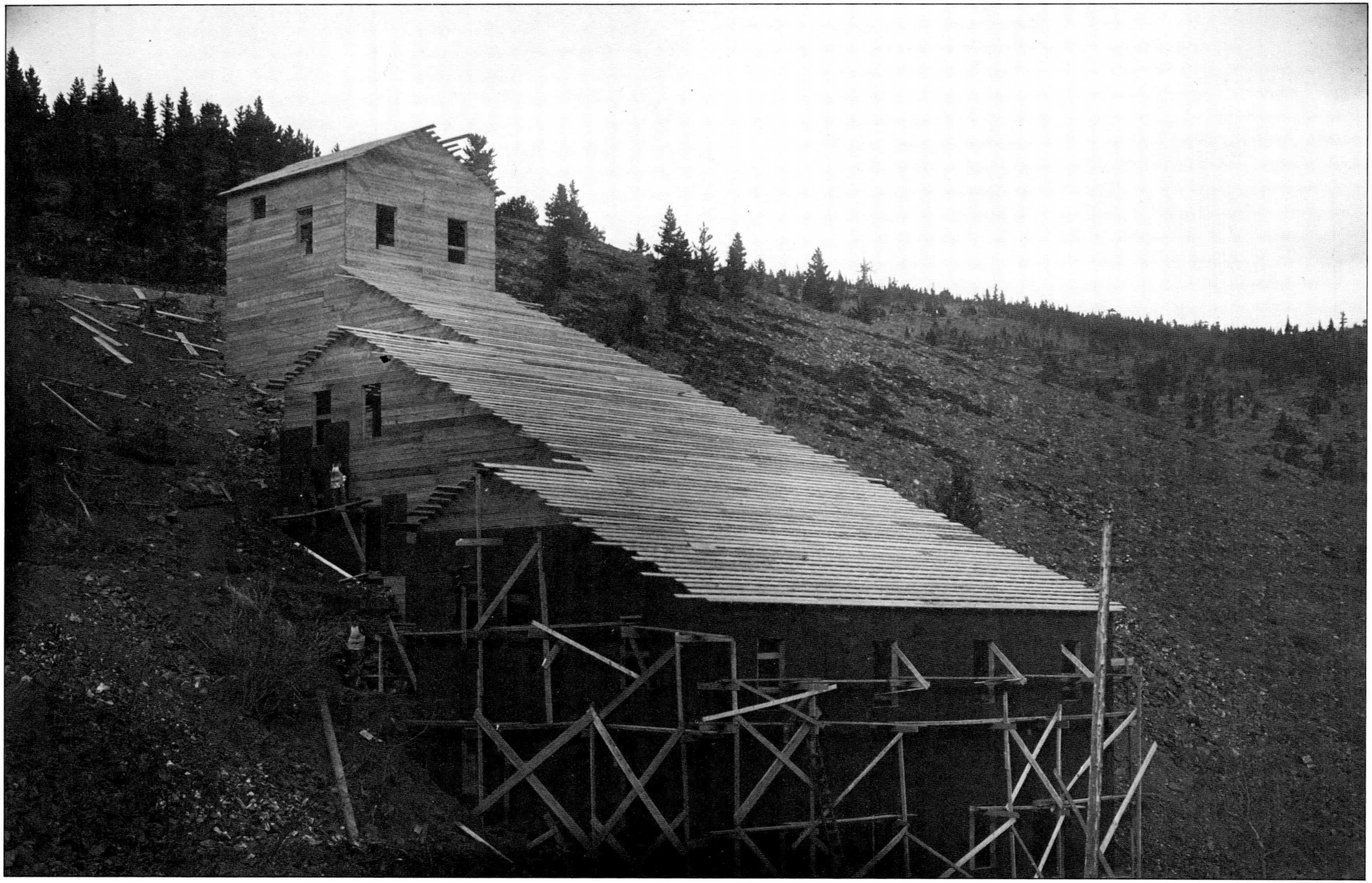

49. Here is an unusual photograph of a mill in the construction stage. The design is typical—to use gravity flow as much as possible. The ore went in at the top, was then crushed, and eventually gold, silver, lead, copper or whatever metal it contained appeared at the bottom. It is possible this may have been a concentration mill where the ore was partially refined and concentrated (e.g., three tons reduced to one of richer value), then shipped on to another plant.

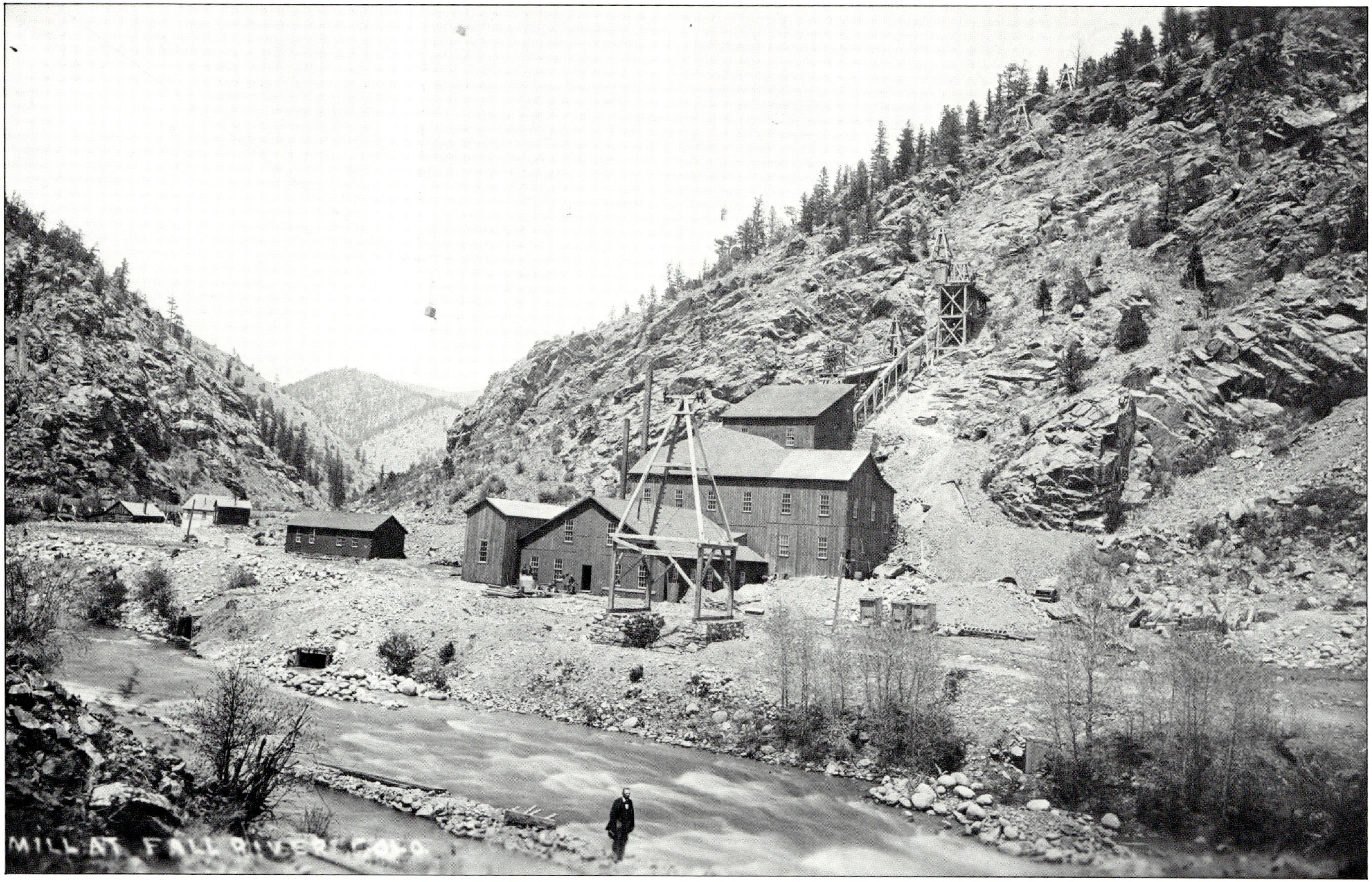

50. This mill at Fall River was served by two trams; one tram bucket can be seen to the left center, as well as a whole series of tram towers stretching off to the right. Trams cut the cost of freighting ore and supplies and were faster than the overland routes. Occasionally a brave miner rode one back to work or down from the mine. Trams connected mine with mill or railroad and could be a significant step in reducing expenses.

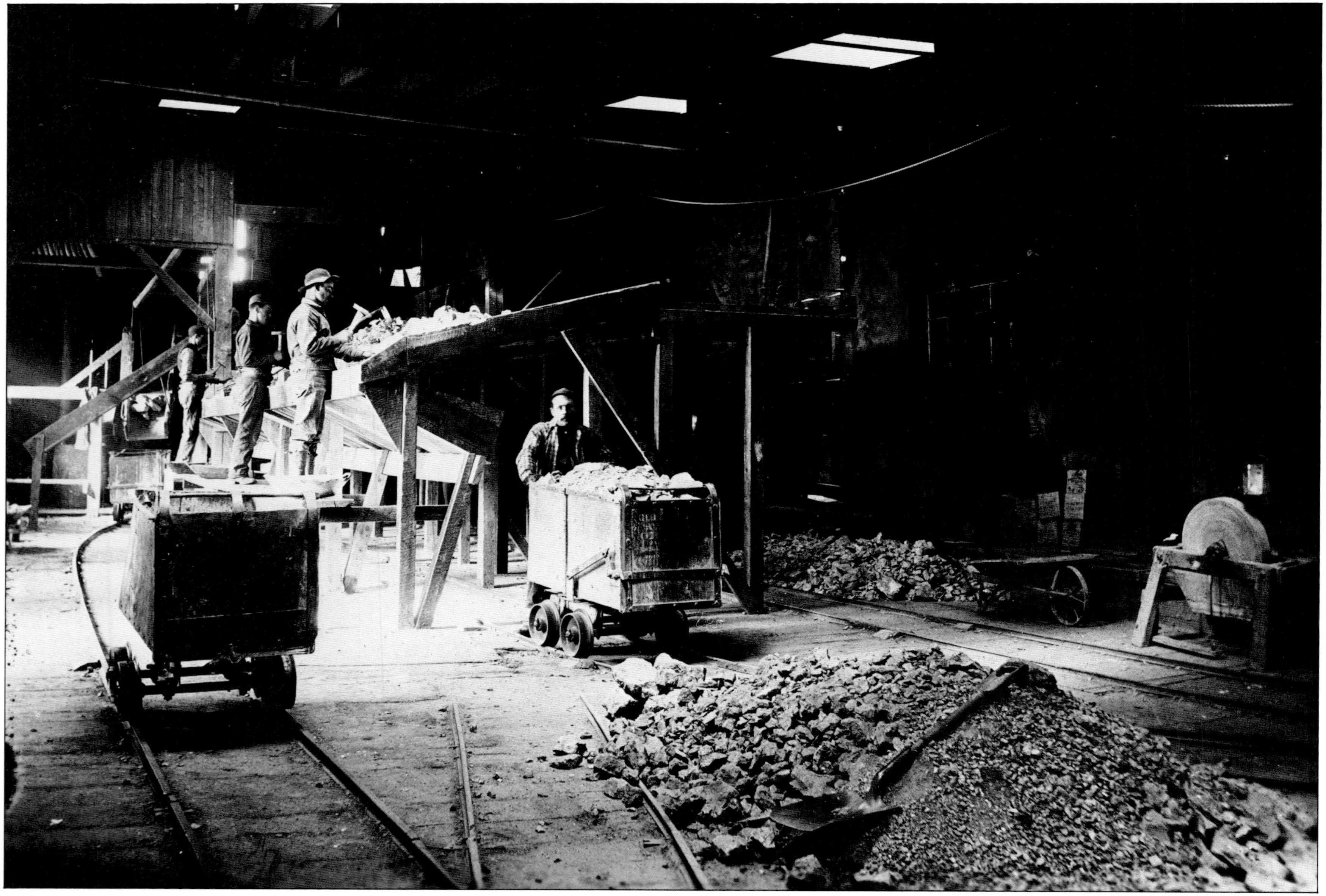

51. In the initial process of dressing the ore at the mill, these men are sorting and picking it over in preparation for reduction. From this point the ore went to be reduced to gold, silver, or other minerals, and the product of the miner's effort deep underground neared fruition.

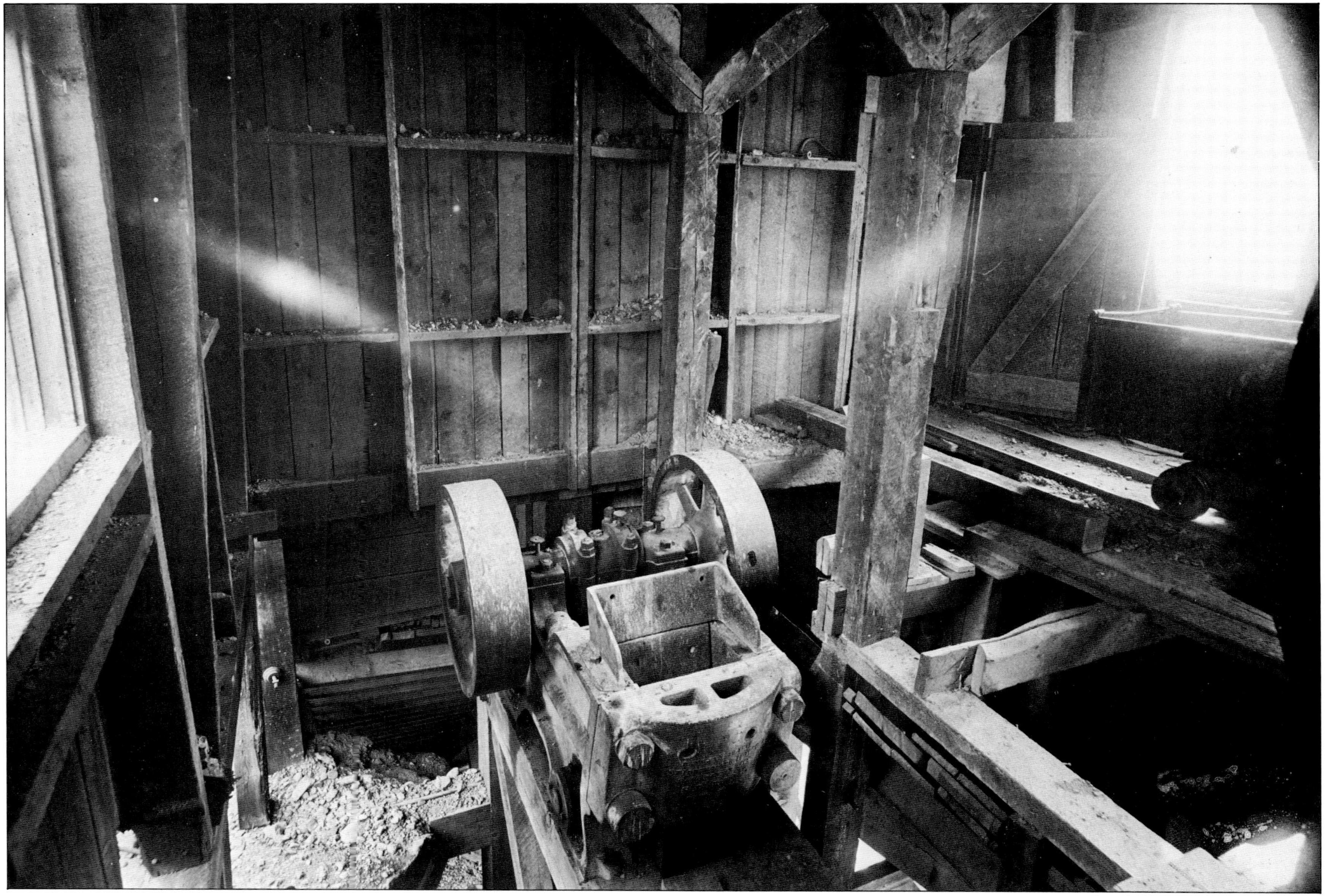

52. Mills were noisy and dusty, starting with the ore being run through this crusher (Krom Patent). After crushing, the ore passed on to be worked further. To withstand the battering, mill timbers had to be sturdy and the buildings were usually multi-storied, stretching along a hillside.

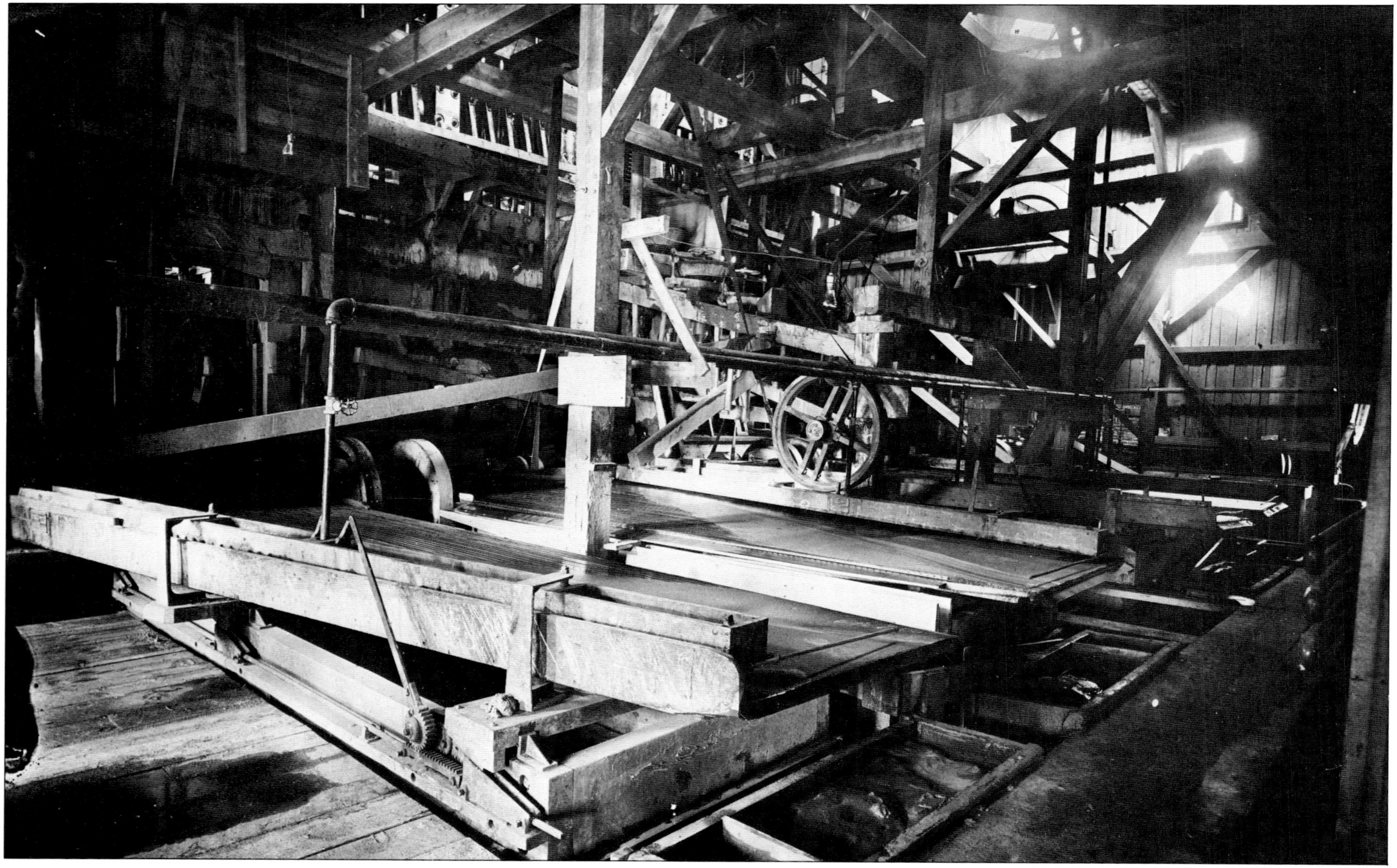

53. By the time the ore reached here, it had been crushed, pulverized by stamps, and combined with mercury and/or chemicals, depending on the process used. The pulp then passed over these concentrating tables, which, with a vibrating motion, maneuvered each mineral behind a predestined riffle bar (depending on mineral weight), and out its exit port. Fascinating to watch, this process was the best of its day. The maze of pulleys, belts, stamps, wires, tables, and pipes was typical.

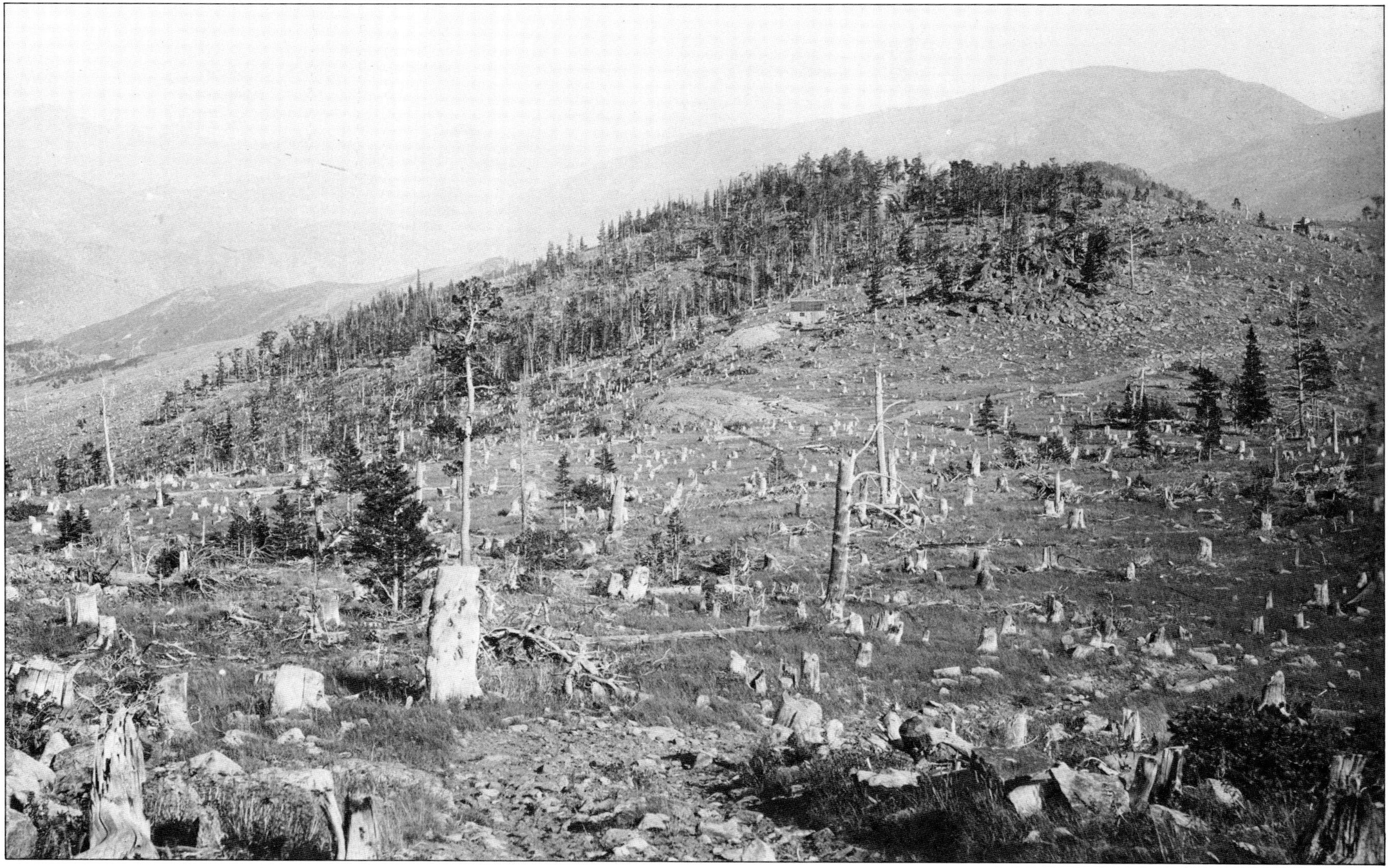

54. In the haste to mine, hillsides were often denuded of trees that went for fuel and mine timbers. Despite a general disregard for conservation and industrial impact on environment (unknown words to the late nineteenth century), there were those people concerned about such waste, because it raised prices and increased expenses. The question was how could it be controlled in an era of rampant free enterprise? This generation did not have an answer.

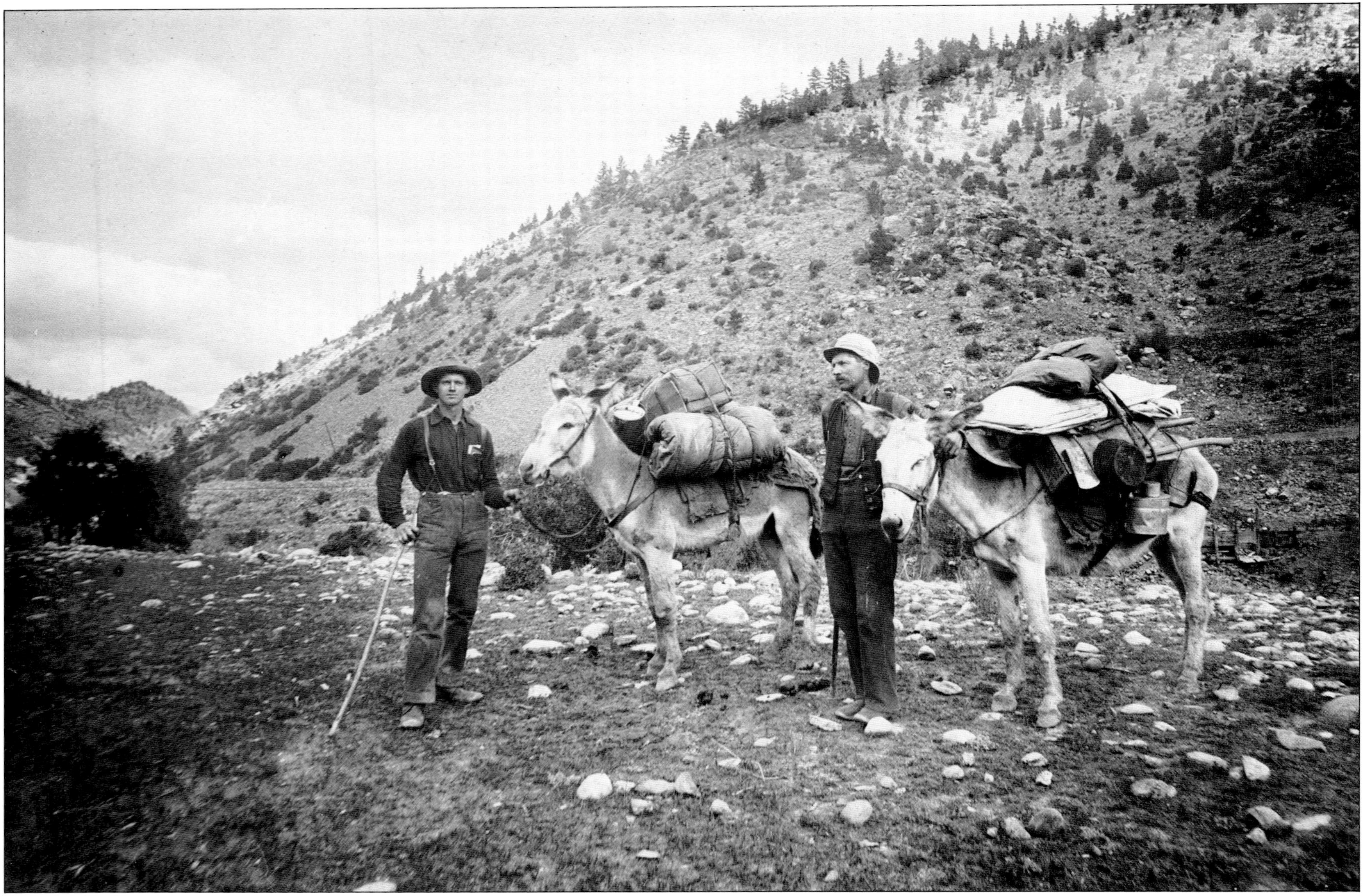

55. And still the cycle goes on. The year could be 1859, 1889, 1909. Prospecting did not stop—somewhere in the hills and mountains of the Clear Creek Valley lay the unfound mother lode, only waiting to be discovered. The swing of a pick, the turn of a shovel, could make them all millionaires. Prospectors and miners through the centuries followed the same "Lorelei's" song.

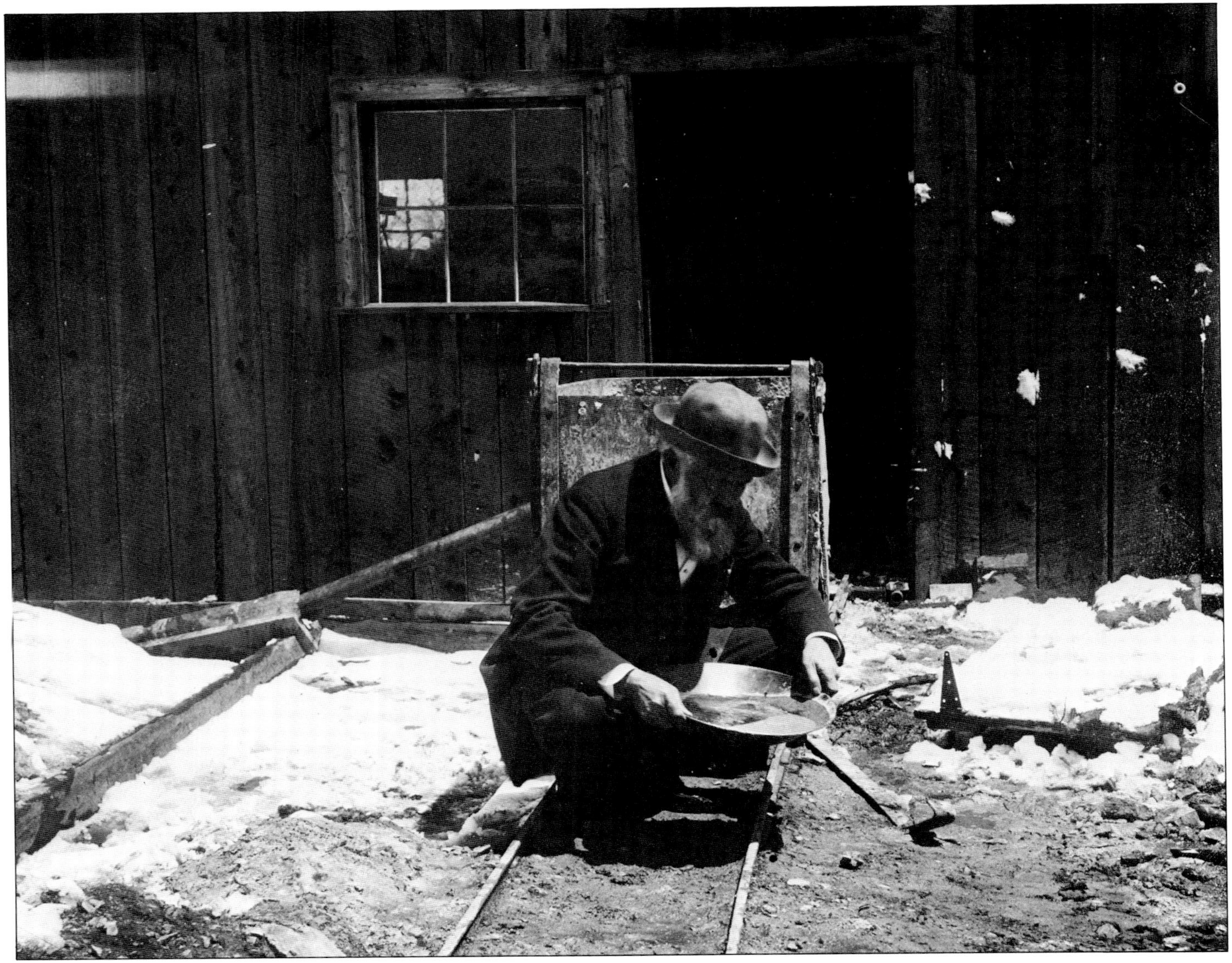

56. Could this be an old fifty-niner showing them how it was done? Actually it is McLean posing for a picture that appeared in a 1902 Chamber of Commerce pamphlet entitled "Health, Wealth & Recreation." Maybe he caught a slight case of gold fever; nevertheless, the photographer and mining had grown old together. Thanks to McLean's work, the "substance did not perish," as William Byers had worried over forty years earlier.

Bibliographical Essay

For the person who wishes to pursue further the subject of mining or the Clear Creek Valley, some reading treats lie in wait. Fitting both into the general state history are Robert Athearn's *The Coloradans* (1976), Carl Abbott's *Colorado* (1976), Marshall Sprague's *Colorado* (1976), and Ubbelohde, Benson and Smith's *A Colorado History* (1976). These are all recent, readable, and provide varying views of the state.

No general history of the Clear Creek Valley has been written since the *History of Clear Creek and Boulder Valleys, Colorado* (1880). The place to start is not with this weighty tome, however. First read one of the finest books to appear about Colorado in many years, James Grafton Rogers' *My Rocky Mountain Valley* (1968).

It is impossible now to relive the nineteenth century, except through photographs and literature. For those who want to know what it was like, the following are recommended: Isabella Bird, *A Lady's Life in the Rocky Mountains* (1873); Samuel Bowles, *The Switzerland of America* (1869); Frank C. Young, *Echoes from Arcadia* (1903); and Henry Villard, *The Past and Present of the Pike's Peak Gold Regions* (1860). No satisfactory history, one which takes their stories from the opening mining era into the twentieth century, of either Central City or Georgetown, has been written. Central's saga has been recounted in numerous books and pamphlets, most closing sometime before 1876. John Horner's *Silver Town* (1950) covers Georgetown into the 1870s, primarily through newspaper stories. Perhaps the best place to begin is with Muriel Wolle's captivating *Stampede to Timberline* (several editions, 1949-74), a volume full of first-hand accounts, bits of history, and drawings by the author. Ethel Gillette's *Idaho Springs: Saratoga of the Rockies* (1978) had potential but did not pan out, and is the best of only a slim offering. *The Colorado Magazine* has a variety of articles and should not be overlooked by the reader who is seeking material on the valley or the industry.

Three contemporary authors of the period describe mining in the 1860s and 1870s: Ovando Hollister, *The Mines of Colorado* (1867), Frank Fossett, *Colorado* (several editions, 1876-1880), and Samuel Cushman, *The Mines of Clear Creek County* (1876). For the more general view, see Wolle or Duane A. Smith, *Colorado Mining* (1977).

For those intrigued by photography, Terry Mangan's *Colorado on Glass* (1975) and William H. Jackson's *Time Exposure* (1940) are excellent. Jackson also has an article in *The Colorado Magazine*, March 1926. Robert Taft's *Photography and the American Scene* (1938) presents a general view.

It is the desire of the authors that this volume will whet the reader's appetite for the history of Colorado. If that has been done, then success will crown our efforts, because the fascinating, wonderful world of Colorado history awaits those who enter it with curiosity and enthusiasm.

Index